Green Energy and Technology

More information about this series at http://www.springer.com/series/8059

Alejandro A. Franco · Marie Liesse Doublet
Wolfgang G. Bessler

Editors

Physical Multiscale Modeling and Numerical Simulation of Electrochemical Devices for Energy Conversion and Storage

From Theory to Engineering to Practice

Springer

Editors
Alejandro A. Franco
Laboratoire de Réactivité et Chimie des
  Solides
Université de Picardie Jules Verne and
  CNRS
Amiens
France

Wolfgang G. Bessler
Offenburg University of Applied Sciences
Offenburg
Germany

Marie Liesse Doublet
Institut Charles Gerhardt CNRS
Université Montpellier
Montpellier
France

ISSN 1865-3529   ISSN 1865-3537   (electronic)
Green Energy and Technology
ISBN 978-1-4471-7066-2   ISBN 978-1-4471-5677-2 (eBook)
DOI 10.1007/978-1-4471-5677-2

Printed on acid-free paper

Springer-Verlag London Ltd. is part of Springer Science+Business Media (www.springer.com)

Preface

World-scale challenges such as climate change, depletion of fossil resources, and the fast increasing energy demand has significantly boosted the R&D on alternative devices for energy conversion and storage. In this context, emerging technologies such as fuel cells and batteries are called to play an important role in any sustainable scenario. However, the successful large-scale implementation of these devices in realistic applications is subjected to numerous constraints in terms of cost, efficiency, durability, safety and impact on the environment. Precise design of cells and stacks is then required, and production cost constraints drive more and more the R&D to go beyond trial-error approaches: the use of numerical simulation and mathematical modeling arises as a natural approach to deal with the design optimization problem.

Since almost 60 years numerous mathematical models of fuel cells and batteries have been reported showing powerful capabilities for *in silico* studies of a large diversity of mechanisms and processes. These models are generally devoted to link the chemical and microstructural properties of materials and components with their macroscopic efficiency. In combination with dedicated experiments, they can potentially provide tremendous progress in designing and optimizing the next-generation cells. The available spectrum of approaches already available is wide: quantum mechanics, nonequilibrium thermodynamics, Monte Carlo and molecular dynamics methods, continuum modeling, and more recently, multiscale and/or multiparadigm models connecting multiple simulation techniques and describing the interplay of mechanisms at multiple spatiotemporal scales.

Through several comprehensive chapters written by recognized scientists in the field, we aim at reviewing the latest progresses in the development and deploy of innovative physical modeling methods and numerical simulations to better understand, rationalize, and predict the electrochemical mechanisms involved in electrochemical devices for energy storage and conversion. Concepts, methodologies, and approaches connecting *ab initio* with micro, meso, and macroscale modeling of the components and cells are revisited, jointly with appropriate illustrations and application examples. Major remaining scientific challenges are also discussed. We hope this book will provide an interesting support to students, researchers and engineers from the industry and academic communities.

Alejandro A. Franco
Marie Liesse Doublet
Wolfgang G. Bessler

Contents

Atomistic Modeling of Electrode Materials for Li-Ion Batteries: From Bulk to Interfaces

Matthieu Saubanère, Jean-Sébastien Filhol and Marie-Liesse Doublet

Abstract In the field of energy materials, the computational modeling of electrochemical devices such as fuel cells, rechargeable batteries, photovoltaic cells, or photo-batteries that combine energy conversion and storage represent a great challenge for theoreticians. Given the wide variety of issues related to the modeling of each of these devices, this chapter is restricted to the study of rechargeable batteries (accumulators) and, more particularly, Li-ion batteries. The aim of this chapter is to emphasize some of the key problems related to the theoretical and computational treatment of these complex systems and to present some of the state-of-the-art computational techniques and methodologies being developed in this area to meet one of the greatest challenges of our century in terms of energy storage.

1 Introduction

Despite the apparent simplicity of the operation of a Li-ion battery, the electrochemical mechanisms involved at the bulk scale and at the interfaces between the electrodes and the electrolyte often rely on complex physical and chemical processes. These processes may occur at different time- and length scales involving static (electromotive force, resistances) and dynamic (charge transfer, mass transport) variables. Addressing these mechanisms simultaneously is intractable at the highest atomistic level of theory, i.e., the ab initio level. Continuum models exist to address battery operation as a whole and under operating conditions of pressure and temperature that rely on phenomenological approaches, e.g., equivalent circuit models [1] or electrochemical kinetic models [2]. Based on the Butler–Volmer equation with varying degrees of sophistication depending on the number of physicochemical effects described, these methods represent a class of "top–down" approaches. They have already benefited the optimization of battery performances

M. Saubanère · J.-S. Filhol · M.-L. Doublet (✉)
Insitut Charles GERHARDT, CNRS and Université de Montpellier, Montpellier, France
e-mail: Marie-Liesse.Doublet@umontpellier.fr

© Springer-Verlag London 2016   1
A.A. Franco et al. (eds.), *Physical Multiscale Modeling and Numerical Simulation of Electrochemical Devices for Energy Conversion and Storage*, Green Energy and Technology, DOI 10.1007/978-1-4471-5677-2_1

[3–6] but are not strictly predictive. In particular, they do not correlate the performance of a battery to the intrinsic properties of electrode materials as no data specific to the chemical/electronic nature of these materials are included.

To build more predictive multiscale models, it is essential to assess quantitative parameters that take into account not only the structural and electronic properties of the electrode materials, but also their reactivity versus lithium (activity, distribution, etc.). At the most local scale (microscopic), first-principles computational methods provide valuable quantitative data, either thermodynamic (electrochemical potentials and equilibrium structures of the electrode material) or kinetic (ion diffusion barriers, interface migration) that can be used as input parameters in models of higher time- and length scales. Among them, Kinetic Monte Carlo simulations, [7] nonequilibrium thermodynamic models [8, 9], or phase-field models [10] intend to simulate time evolution of some elementary processes—whose rates have been previously investigated at the ab initio level (atomic scale)—in order to determine their respective impact on the structure of the electrodes at the nanoscopic scale. These approaches are the so-called bottom–up approaches. They have already proven to be powerful in the field of fuel cells [11] and are currently under development in the field of Li-ion batteries [12, 13]. They represent a holy grail for theoreticians involved in this field, not only to ensure the transferability of parameters from one scale to another but also to solve the nonlinearly coupled equations of ion and electron transport which govern the macroscopic behavior and lifetime of electrochemical cells.

Pending this ultimate method, it can be useful to combine the "bottom–up" and "top–down" approaches to study the electrochemical battery performance by taking advantage of (i) the limited computational cost of phenomenological approaches and (ii) the accurate description of materials and interfaces provided by atomistic quantum methods. These approaches are promising but still very limited in the field of Li-ion batteries, not only due to the large number of parameters to be extracted but also to the methodological and numerical locks associated with ab initio methods for extracting these parameters. Thus, in this area, the vast majority of computational scientists focus their studies on one element of the battery (electrode or electrolyte) in order to optimize their performance with respect to strict industrial specifications. Focusing on battery materials, the candidates have to be safe, cheap, and environmentally friendly and must meet several criteria such as high-energy density, good rate capability, and long-term cycling life. Besides the economic and ecologic aspects on which chemists can act to meet the industrial specifications, fundamental chemistry can also be used to improve the electrochemical performance of electrode materials. To that aim, first-principles quantum methods may provide solid-state chemists with a powerful tool to reproduce, understand, or predict material properties. On the one hand, they give access to thermodynamic or kinetic quantities such as equilibrium structures and energies or energy barriers that are further used to interpret the microscopic parameters influencing the structural and electronic behavior of a material. In that sense, they are useful to decide whether or not a given material is suitable for the application of interest and whether or not chemical functionalizations or appropriate engineering could improve the material performance. On the other hand, these studies aim at parameterizing multiscale models with quantitative thermodynamic and kinetic inputs which are sometimes inaccessible from experiments. In this way, computational chemists invested in this area contribute—even unconsciously—to feeding the virtual database of input parameters required for the next generation of multiscale "bottom–up" models.

In this chapter, we will give a non-exhaustive review of what first-principles calculations can bring to the understanding of material performance and to the extraction of quantitative thermodynamic and kinetic parameters, with a special focus on the methodological and numerical remaining locks. Basic thermodynamics are first used to describe the energetics of electrochemical reactions underlying the battery operation. Then, a brief review of first-principles approaches to condensed matter is given, along with technical aspects/locks associated with the computation of the thermodynamic quantities of interest. The last section is devoted to the perspectives.

2　Macroscopic Picture of an Electrochemical Reaction

The **voltage** delivered by a rechargeable Li-ion battery is a key parameter to qualify the device as promising for future applications. It is a thermodynamic quantity that is directly linked to the difference in the electric potentials of the two electrodes constituting the Li-ion cell. During the discharge, two simultaneous redox reactions occur at both electrodes when Li^+ ions and electrons are transferred from the negative electrode (low-potential vs. Li^+/Li^0) to the positive electrode (high-potential vs. Li^+/Li^0). At each step of the reaction x, the equilibrium battery voltage, $V(x)$, is directly linked to the reaction Gibbs energy $\Delta_r G(x)$ through the Nernst equation:

$$\Delta_r G(x) = -nFV(x) \tag{1}$$

where F is the Faraday constant and n is the number of charge transported through the electrolyte by the exchanged (Li) ions. In order to specifically check the performance of a given electrode material, hereafter denoted $\mathcal{H}$, chemists usually build half-cells in which the Li-metal reference electrode is used at the negative electrode of the cell. In this way, the battery voltage directly gives the material potential with respect to the Li^+/Li^0 reference potential (-3.04 V vs. NHE). During operation, the electrochemical reactions occurring at both electrodes are

$$\begin{array}{c} Li \rightleftarrows Li^+ + e^- \\ \frac{1}{\varepsilon}Li_x\mathcal{H} + Li^+ + e^- \rightleftarrows \frac{1}{\varepsilon}Li_{x+\epsilon}\mathcal{H} \\ \hline \frac{1}{\varepsilon}Li_x\mathcal{H} + Li \rightleftarrows \frac{1}{\varepsilon}Li_{x+\epsilon}\mathcal{H} \end{array} \tag{R1}$$

In open-circuit conditions (OCV), the total Gibbs energy of the electrochemical system at any given temperature (T) and pressure (p) conditions is

$$G_{\{n_i\},T,p}(x) = \sum_i n_i \mu_i = n_\mathcal{H} \mu_{Li_x\mathcal{H}} + (n_{Li} - x.n_\mathcal{H}) \mu_{Li} \tag{2}$$

where μ_i are the chemical potentials of species i, $n_\mathcal{H}$ and n_{Li} the total number of mole of $\mathcal{H}$ and Li and $x \cdot n_\mathcal{H}$ the reaction extent. The molar free energy of reaction R1 is

$$\Delta_r G = \frac{1}{n_\mathcal{H}} \frac{\partial G(x)}{\partial x} = \frac{\partial \mu_{Li_x\mathcal{H}}}{\partial x} - \mu_{Li} \tag{3}$$

Using the Nernst relation, the electrode potential $V(x)$ with respect to the Li^+/Li^0 reference is then

$$V(x) = -\frac{1}{F} \left\{ \frac{\partial \mu_{Li_x\mathcal{H}}}{\partial x} - \mu_{Li} \right\} = -\frac{1}{n_\mathcal{H} F} \frac{\partial G(x)}{\partial x} \tag{4}$$

From this equation, it is possible to extract the potential variation with lithiation

$$\frac{\partial V(x)}{\partial x} = -\frac{1}{F} \frac{\partial^2 \mu_{Li_x\mathcal{H}}}{\partial x^2} = -\frac{1}{n_\mathcal{H} F} \frac{\partial^2 G(x)}{\partial x^2} \tag{5}$$

At equilibrium, the free energy function of the electrochemical system $G(x)$ must be a convex function of the reaction extent since the electrode achieved at each given composition is assumed to be the most thermodynamically stable electrode. This does not preclude multiphasic electrodes as we will see below. The second derivative of $G(x)$ with respect to x is then positive (or nil), whatever the reaction mechanism (i.e., single- or two-phase process). This implies that the equilibrium material potential $V(x)$ with respect to Li^+/Li^0 decreases (or remains constant) upon lithiation

$$\frac{\partial V(x)}{\partial x} \leq 0 \tag{6}$$

The Gibbs function of the $\{Li_x\mathcal{H} + (n_{Li} - x)Li^0\}$ system, hereafter denoted as $G(Li_x\mathcal{H}/Li)$ is represented in Fig. 1 for $n_\mathcal{H} = 1$ as a function of the Li content x inserted in $Li_x\mathcal{H}$ (green line). When $G(Li_x\mathcal{H}/Li)$ shows a concave shape, i.e., in the composition range $x_1 \leq x \leq x_2$, the electrode is metastable and disproportionates into $\{Li_{x_1}\mathcal{H} + Li_{x_2}\mathcal{H}\}$ to maintain the convexity of $G(x)$ (red line). In that case, the electrochemical mechanism corresponds to a *two-phase process* along which the electrode is a proportional mixture of two single phases of distinct Li compositions. For all other compositions ranges where $G(Li_x\mathcal{H}/Li)$ shows a convex shape, the electrochemical mechanism corresponds to a *single-phase process* along which the electrode is a solid solution in Li composition. The shape of $G(x)$ thus sets the

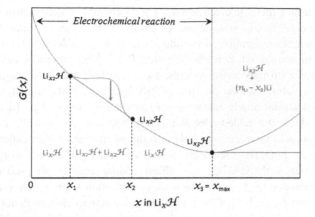

Fig. 1 Evolution of the electrode free energy $G(x)$ along the electrochemical reaction with Li. The *green line* stands for the Gibbs function of the $\{Li_x\mathcal{H} + (n_{Li} - x)Li^0\}$ system in which the $Li_x\mathcal{H}$ phase is a single-phase of composition x. The *red line* indicates the convex hull of $G(x)$, i.e., the pathway followed by the electrode to minimize its free energy. The *gray rectangle* indicates the domain in composition where the electrochemical reaction is energetically no longer achievable (Color figure online)

thermodynamic electrochemical mechanism of the total reaction as well as the intermediate phase compositions which are expected to be stabilized, and therefore observed upon lithium insertion/de-insertion.

The slope of the $G(x)$ function is directly proportional to the potential that must continuously decrease with lithiation. Thus, the composition x_{max} for which the electrochemical potential vanishes sets the end of the electrochemical reaction and therefore the theoretical **capacity** of the electrode material. This quantity is important for an electrode material as it corresponds to the maximal number of Li the material is able to accommodate or release during a cycle of discharge/charge. It is related to the amount of energy stored by the material per mass or volume unit and contributes to the total **energy density** $(\mathcal{E})$ of the material through

$$\mathcal{E} = \frac{F}{\mathcal{M}} \int_{x_{min}}^{x_{max}} V(x)dx \tag{7}$$

where $\mathcal{M}$ is the molar mass of the material.

2.1 Microscopic Picture of an Electrochemical Reaction

As shown in the previous section, the overall reaction corresponds to a variation in the Li chemical composition of the intercalation material, $\mathcal{H}$. The distribution of the added charges (Li^+ ions and electrons) into the host material obviously differs from that of the reference metallic electrode (Li^0) and depends on the electronic structure of the host material, that is, whether the electrons are localized or delocalized in the system. The spatial localization of the added electron (ρ_{e^-}) defines the *redox active center* of the host material, hereafter denoted RAC. High-potential cathode materials are generally strongly ionic systems, e.g., transition metal oxides, phosphates, or sulfates in which the added electron mainly localizes on the transition metal. In contrast, low-potential anode materials are covalent systems, e.g., Li-intercalated graphite, in which the added electron delocalizes on the system. Based on a qualitative approach, Goodenough [14] introduced in 1997 the so-called inductive effect to link the potential variations of a wide variety of transition-metal based electrode materials to the RAC electron affinity. More recently, we used a first-order perturbative approach to show that the ionic contribution to the reaction free energy may also be significant [15]. Given the general electrochemical Reaction R1 we rewrote Eq. (3) as an expansion of the host matrix energy with the added charges

$$\Delta_r G = \Delta E_{Li^+/\mathcal{H}} + \Delta E_{e^-/\mathcal{H}}$$

$$= q_{Li^+} V_{\mathcal{H}}(r_A) + \left(\mu_{RAC}^0 + 2\eta_{RAC} + \int\limits_{RAC} \rho_{e^-}(r) V_{\mathcal{H}}(r) dr \right) \tag{8}$$

In this expression, $V_{\mathcal{H}}(r_A)$ and $V_{\mathcal{H}}(r)$ are the electrostatic fields exerted by the ions of the host matrix at the Li and RAC sites, μ_{RAC}^0 is the chemical potential of the RAC which can be assimilated to the Fermi level of the host material and η_{RAC} is the RAC chemical hardness. The reaction free energy and therefore the voltage amplitude are governed by the electronic reduction of the RAC which is directly linked to its chemical nature (short-range effects) and by the electrostatic modulations associated with the addition of two separated charges (e^- and Li^+) which is directly linked to the crystal structure and polymorphism (long-range effects). Including both the electronic and ionic contributions to the reaction free energy, this approach not only generalizes the qualitative approach of Goodenough [14] but also allows rapid and quantitative assessment of materials potential.

2.2 Beyond the Thermodynamic Equilibrium

Besides the thermodynamic quantities that define whether or not a material is suitable for application, kinetic quantities related to the electrochemical reaction rates are also of great importance for battery performance. In particular, the ionic and electron transport into the host matrix and/or at the interfaces between the active material and the electrolyte (Li^+ transport) or the current collector (electron transport) may lead to kinetic limitations, which in turn alter the electrochemical mechanisms. These limitations generally induce electrode polarization effects (gradient of Li composition and/or local electric fields) that directly translate into

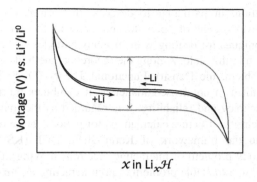

Fig. 2 Effect of electrode polarization (*red*) on the voltage profile of a $Li_x\mathcal{H}/Li$ half-cell in operating charge (−Li) and discharge (+Li) conditions with respect to the thermodynamic equilibrium (*black*) (Color figure online)

different working voltages between the charge and the discharge. As shown in Fig. 2, the electrode follows a kinetically activated pathway that is no longer dictated by the convex $G(x)$ function and is associated with different potentials in charge (delithiation) and discharge (lithiation). Such kinetic limitations obviously affect the efficiency of the device as the energy required to recharge the battery is higher than the energy delivered in discharge. To overcome this issue, chemists often play with appropriate engineering of the electrode (particle size, carbon additives, and binders) to increase the electric contacts and ionic diffusion rates inside the particles and at grain boundaries. Often beneficial to the battery performance, the impact of these formulations is, however, not trivial. Indeed, the decrease in the particle size may promote some interfacial electrochemical effects that are not well understood, yet are crucial in the prediction and understanding of the overall reaction mechanism. It is therefore important to account for these phenomena when studying the electrochemical properties of battery materials. This supposes the development of new methodologies to capture and include these effects in the response property of electrode material under electrochemical conditions.

2.3 First-Principles Approach to Condensed Matter

The first-principles approach to condensed matter consists in starting from what we know about a material, i.e., its chemical composition and to calculate its energy and properties. The interactions between atoms, such as chemical bonding, are determined by the interactions of their constituent nuclei and electrons. In the Born–Oppenheimer approximation, the electrons are the particles setting the many-body problem and their behavior is governed by basic quantum mechanics. What makes first-principles calculations difficult is mainly the size of the problem in terms of a numerical formulation, in particular to describe the complex issue of electron repulsions. The development of accurate and efficient theoretical methods and computational techniques for dealing with so many particles is therefore central to the ongoing research in this field. Among the theoretical methods available to solve the many-body problem, the Density Functional Theory (DFT) [16] is the most widely used formalism in condensed matter as it combines numerical efficiency with acceptable accuracy and reliability. It is regarded as the main computational tool to perform electronic structure calculations for periodic systems with a realistic complexity. Within the framework of Kohn–Sham DFT (KS DFT), [17] the intractable many-body problem of interacting electrons moving in a static external potential is reduced to a tractable problem of noninteracting electrons moving in an effective potential:

$$\left\{ -\frac{\hbar^2}{2m} + v_{\text{eff}}(\mathbf{r}) \right\} \phi_i(\mathbf{r}) = \varepsilon_i \phi_i(\mathbf{r}) \tag{9}$$

The effective potential $v_{\text{eff}}(\mathbf{r})$ includes the electrostatic external potential set by the atomic coordinates (nuclei) and the Coulomb interactions between electrons modulated by many-body effects:

$$v_{\text{eff}}(\mathbf{r}) = v_{\text{ext}}(\mathbf{r}) + e^2 \int \frac{\rho(\mathbf{r}')}{|\mathbf{r} - \mathbf{r}'|} d\mathbf{r}' + \frac{\partial E_{\text{xc}}[\rho]}{\partial \rho(\mathbf{r})} \tag{10}$$

The $E_{\text{xc}}[\rho]$ term is the so-called exchange-correlation energy term. It includes corrections to the kinetic and Coulomb energies but its exact expression is unknown. Various flavors of energy functionals were developed so far to reach an accurate description of this energy functional, among which the most popular are the Local Density Approximation (LDA) [18] and the Generalized Gradient Approximation (GGA) [19, 20]. Although very powerful in describing the ground-state structure, energy and properties of a wide variety of materials ranging from metals to large-gap semiconductors, these functionals are unable to capture the physics of strongly correlated electrons or weak interactions. In particular, transition metal oxides such as those known as Mott–Hubbard insulators (NiO) are predicted to be metallic in the LDA or GGA approximations due to the so-called self-interaction error (SIE), which tends to over-delocalize the electrons along the chemical bonds. In these cases, more sophisticated functionals, e.g., hybrid (DFT/HF) [21–23], range-separated (HSE0) [24], or Hubbard-corrected (DFT+U) [25–27] functionals are required to properly describe the energetics and electronic structure of these materials. As we will see in the following, these functionals rely on adjustable parameters whose reliability needs to be evaluated before use. The DFT/HF functionals incorporate a portion of the exact Hartree–Fock exchange energy in the E_{xc} term to balance the overestimated DFT exchange energy. In condensed matter, the most popular hybrid functional is the PBE0 (Perdew–Burke–Ernzerhof) [23] which mixes the Hartree–Fock and DFT exchange energics in a 1:3 ratio:

$$E_{xc}^{PBE0} = \frac{1}{4} E_x^{HF} + \frac{3}{4} E_x^{PBE0} + E_c^{PBE0} \tag{11}$$

The HSE0 (Heyd–Scuseria–Ernzerhof) formalism introduces another adjustable parameter λ to discriminate the regions in space where the HF/DFT mixing is needed:

$$E_{xc}^{\lambda PBE0} = a E_x^{HF,SR}(\lambda) + (1 - a) E_x^{PBE0,SR}(\lambda) + E_x^{PBE0,LR}(\lambda) + E_c^{PBE0} \tag{12}$$

where the mixing parameter a is 1/4 for the HSE06 functional and SR and LR stand for short- and long-range interaction regions. The idea behind the DFT+U formalism is physically different from that of hybrids or range-separated functionals. It consists in describing the strongly correlated (localized) electronic states of the system with a Hubbard-like model, and to treat the rest of valence electrons with standard DFT functionals. Practically, an effective on-site Coulomb U_{eff} (also adjustable) is added to the subset of strongly correlated orbitals i (e.g., d- or f-orbitals) in order to penalize their partial occupation. The simplest expression of the DFT+U energy functional is the rotationally invariant approximation of Dudarev et al. [27] and is given by

$$E^{DFT+U}[\rho, \{n^{\sigma}\}] = E^{DFT}[\rho] + \{E^{Hub}[n^{i\sigma}] - E_{dc}[n^{i\sigma}]\} \tag{13}$$

where ρ is the total density and $n^{i\sigma}$ the spin occupation matrix of the strongly correlated orbitals i. While the U_{eff} parameter can be extracted from a self-consistent procedure based on a linear-response approach [28], it is most widely used as an adjustable parameter.

How efficient are first-principles calculations to investigate the electrochemical properties of complex electrode materials? A literature survey is sufficient to realize that the pioneer work of Ceder and coworkers in 1997 on first-principles predictions of electrode potential [29] has opened a tremendous field of investigation for theoreticians. Ever since, the number of computational studies devoted to the electrochemical properties of electrode materials keeps on increasing every year. Besides, new methodologies emerge to (i) improve the accuracy of the calculations, (ii) describe more finely the electrochemical mechanisms, and (iii) introduce experimental reality such as particle size, pressure and temperature conditions, and so on. In the following, we give a non-exhaustive review of the methodologies used to link the thermodynamic equations of Sect. 2 to first-principles calculations. We distinguish systems for which the electrochemical properties are governed by the bulk phase (Sect. 3) from those for which interface electrochemistry is dominant (Sect. 4).

3 Modelization of Bulk Materials

As shown in the previous section, the overall electrochemical mechanism whereby a material reacts with Li can be fully predicted from the finite-temperature phase diagram of Fig. 1. A first step is then to evaluate the material free energy $G(Li_x\mathcal{H})$ which depends on the material chemical composition and crystal structure, and to a lower extent on external conditions of temperature, pressure/stress or electric field. In the limit of large particle sizes, the pressure/stress and electrical field contributions to the material free energy can be neglected. Indeed, the local stress arising from volume and structural changes is mostly released by dislocations and no significant electric fields are expected in bulk at equilibrium due to an efficient mutual screening of the added charges (e^- and Li^+). As we will see later in the next section, this is no longer true for surfaces due to the electrical double layer (EDL). Since typical working pressures (~ 1 bar) have no significant effect on condensed matter, what finally governs the free energy of a bulk material is its chemical composition, its crystal structure, and the external temperature. Therefore, assessing the material Gibbs energy at each Li composition and finite temperature should be sufficient to determine the redox mechanism (i.e., the number of electrochemical processes and the related intermediate electrodes) and the thermodynamic properties of the system (potential, capacity, thermal stability) at equilibrium. When necessary, kinetic properties such as Li diffusion pathways and their associated energy barriers can also be investigated in a second step. This will be briefly discussed in Sect. 3.3.

Focusing on thermodynamics, the computation of $G(Li_x\mathcal{H})$ requires a multi-step procedure: (i) first, equilibrium structures and energies need to be computed from $T = 0$ K periodic calculations within a reasonable numerical accuracy for various Li-compositions and Li-distributions; (ii) then finite temperature effects need to be included to check their impact onto the material energy and to figure out the reaction pathway(s); (iii) finally, the thermodynamic (potential, capacity) and kinetic (energy barrier for Li-diffusion) properties can be evaluated.

3.1 Equilibrium Crystal Structures

In order to reproduce the structural modifications a material is susceptible to undergo upon lithiation/delithiation, an accurate description of the material equilibrium structure is required at each given x composition. To that aim, routine procedures based on the Hellmann–Feynman scheme (atomic force calculation) [30] are used to explore the material potential energy surface (PES) and search for its energy minimum (i.e., most thermodynamically stable structure). They are based on deterministic schemes which guide the system to the closest local minimum of the PES by following the surface downhill from an arbitrary starting point. These procedures, known as "Steepest Descent" and "Conjugate Gradient" procedures are implemented in most of the commercialized DFT codes devoted to periodic systems and lead generally to excellent agreement with experimental structures. However, they are known to be local energy minimization procedures and cannot guarantee reaching the global energy minimum of high-dimensional PES, in particular when the guess structure (input of the calculation) is far away from the equilibrium structure or when different polymorphs exist at a given chemical composition. For such complex search, more sophisticated methods are required to ensure that the material PES is more extensively sampled.

To solve the global space-group optimization problem, various algorithms can be used to sample the material PES and search for its global minimum. **Genetic algorithms** such as those implemented in the USPEX program package code for structural prediction [31] mimic the Darwin principle to create new "child" structures from an initial population of randomly generated "parent" structures. The "child" structure is appropriately designed to preserve some of the structural properties of the two candidates chosen as the "parent" structures in the initial population. They evolve through an evolutional algorithm that relies on mating and mutation operations, and the validity of new "child" structure is evaluated against various constraints and local minimization procedures.

A preconditioning step can be used to select the initial population. A systematic way to do this is to formalize and codify the historical knowledge on materials structures and properties. This idea is the basis of the *data-mining-driven quantum mechanics* method proposed in 2006 by Ceder et al. [32] for structural prediction. The data-mining method is based on a **Bayesian probabilistic algorithm** and appeals to what we know from decades of experimental and computational works, in order to select the most probable candidate structures to be further evaluated by quantum mechanics (DFT). This method is the one implemented in the Material Genome Project of the M.I.T. and is extensively used in the field of Li-ion batteries [33].

More **stochastic algorithms** such as the Simulated Annealing [34] (SA) or Basin-Hopping [35] (BH) procedures also provide global optimization (see Fig. 3). In contrast to the two previous ones, these procedures start with one arbitrary structure configuration and generate new structures in randomly displacing the atoms.

In the SA scheme, the temperature is used to avoid the system to be trapped in a local minimum. Starting with a high temperature, the system is progressively cooled down to ultimately reach the global minimum of the PES. This method is generally well adapted to PES with not too high-energy barriers between local minima. In the BH scheme, the temperature is no longer used as a parameter to enable the system to cross the energy barriers but is kept constant along the procedure. Instead, a local structural relaxation is performed for each newly generated structure. The change in the total energy (ΔE) is used to accept or discard the new

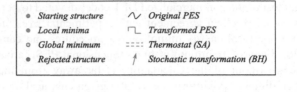

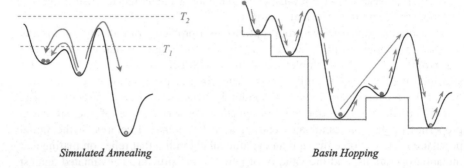

Fig. 3 Schematic representation of the Simulated Annealing (*left*) and Basin-Hopping (*right*) procedures

generated structures[1] via a Metropolis algorithm. In this way, the PES is transformed into a canonical ensemble of local minima from which thermodynamic properties can be derived.

> In the field of Li-ion batteries, when the electrode material crystal structure is known from experiments (X-ray or neutron diffraction) either in its Li-rich or Li-poor composition and remains mostly unchanged upon Li insertion/de-insertion, local energy minimization procedures might be sufficient to reach accurate equilibrium structures for every Li-composition achieved during the electrochemical process. Nevertheless, it is important to keep in mind that this result does not apply for any type of electrode material and that global minimization procedures must be used anytime an important structure change (or phase transition) is expected for the electrode material upon Li-insertion/de-insertion.

---

[1]The acceptance criterium is such that (i) if $\Delta E \leq 0$, the new structure is used as the new guess structure or (ii) if $\Delta E > 0$ the new structure is assigned to a probability $P(E) = e^{-(\Delta E/k_B T)}$, thus leading to a canonical ensemble of atomic configurations at T.

3.2 Finite Temperature Effects

In battery materials, the host matrix chemical bonds and cell volume can be significantly modified by the insertion of Li. Moreover, the change in Li-composition can generate statistical disorder over the cationic sites of the host matrix.[2] At finite temperature, the thermal energy due to lattice vibrations and/or Li-ordering can be important enough to influence the electrochemical mechanism of Li-insertion and must therefore be evaluated. Ab initio molecular dynamics (AIMD) [36] is, in principle, the method of choice to account for these thermal effects. It is however computationally too expensive to reach the timescale often needed for the thermodynamic convergence of material properties. A cost-effective alternative to evaluate these thermal contributions to the free energy is to combine static ($T = 0$ K) first-principles calculations with statistical physics extrapolation to finite temperature.

In condensed matter, the thermal energy arising from lattice vibrations contributes both to the system enthalpy and to the system entropy, while the thermal energy arising from Li-ordering contributes only to the system entropy. In standard pressure conditions, the composition and temperature-dependent free energy then decomposes into

$$
\begin{aligned}
G_{n_i,T,p^0} &= H_{n_i,T,p^0} - TS_{n_i,T,p^0} \\
&= H^0_{n_i,T=0\,\mathrm{K}} + H^{\mathrm{vib}}_{n_i,T} - T\left\{ S^{\mathrm{vib}}_{n_i,T} + S^{\mathrm{conf}}_{n_i,T} \right\}
\end{aligned}
\tag{14}
$$

$H^0_{n_i,T=0\,\mathrm{K}}$ is assimilated to the ground-state electronic energy obtained from first-principles $T = 0$ K calculations. In the harmonic limit, the **thermal vibration contribution to enthalpy** H^{vib} is found by integrating the energy of lattice vibrations over the distribution of frequencies

$$
\begin{aligned}
H_{\mathrm{vib}}(T) &= \int\limits_0^{+\infty} \left\{ \frac{1}{2}\hbar\omega + \frac{\hbar\omega}{(e^{\hbar\omega/k_B T} - 1)} \right\} g(\omega)\,d\omega \\
&= E_{\mathrm{ZPE}} + E_{\mathrm{phon}},
\end{aligned}
\tag{15}
$$

where $g(\omega), k_B, T$ are the phonon density of states at a given frequency ω, the Boltzmann constant and the temperature, respectively. At low temperature ($\hbar\omega \gg k_B T$) thermal vibrations are dominated by the so-called **zero-point energy**, E_{ZPE}, which arises from nuclear quantum effects. E_{ZPE} corresponds to the vibrational ground-state of the 3N vibrations of the system. This term is dominated by high-energy phonons and may be important for light atoms such as H and Li. At high temperature ($\hbar\omega \ll k_B T$) the thermal vibration contributions to enthalpy

---

[2]We should note here that not only the Li^+ ions but also the constituting elements of the host matrix can show some disorder in the structure.

are dominated by the phonon energy E_{phon} which shows an asymptotic behavior to $3N\,k_B T$.

The phonon density of states can be computed from first principles but becomes rapidly prohibitive for large systems, especially when long-range electrostatic coupling occurs in the structure. To reduce the computational cost, a usually good approximation for large systems is to restrict the phonon band calculation at the Γ point (Brillouin zone center) and to extract the phonon frequencies that are used to compute $H_{vib}(T)$. In polar crystals, the macroscopic electric fields induced by the interactions of charged planes are associated with long wave (low-energy) longitudinal optical phonons and are responsible for the well-known LO–TO splitting phenomenon [37]. Accounting for this LO–TO coupling in a Γ-point calculation then requires the use of extended super cells to properly describe the vibration properties of the system [38].

In electrode materials, the change in Li-composition can significantly modify the phonon band structure and therefore the phonon energy. Both E_{ZPE} and E_{phon} are therefore affected and should, in principle, be computed. Nevertheless, they are often neglected as they are far smaller in magnitude than the electrochemical energies.

Also important is the entropy contribution to the free energy. For any given Li-composition, the two main contributions to entropy are the **vibration** term, i.e., how disordered are the phonons in the accessible vibration states of the system and the **configuration** term, i.e., how disordered are the Li$^+$/Li-vacancies over the host material cationic sites. The electronic entropy can be neglected at room temperature. In the harmonic limit, the vibration entropy is estimated by statistical physics as

$$S_{vib}(T) = k_B \int\limits_{0}^{+\infty} g(\omega) \left\{ \frac{\hbar\omega}{k_B T (e^{\hbar\omega/k_B T} - 1)} - \ln(1 - e^{-\hbar\omega/k_B T}) \right\} d\omega \qquad (16)$$

Similar to H^{vib}, S^{vib} can be evaluated from first-principles phonons (vibration modes) calculations but is again computationally expensive while it contributes to the system free energy to a few tens of meV per formula unit.

In contrast to vibration entropy, configuration entropy has no general expression and cannot be straightforwardly extrapolated from one first-principles calculations. In the canonical ensemble defining the full configuration space of the system at a given Li-composition, it is written as

$$S_{conf}(T) = -k_B \sum_i P_i \ln P_i \qquad (17)$$

where P_i is the probability of occurrence of configuration i with energy E_i:

$$P_i = \frac{1}{Z} \exp^{-(E_i/k_B T)} \tag{18}$$

Configuration entropy depends on the canonical partition function Z which describes the statistical properties of the system in thermodynamic equilibrium:

$$Z = \sum_i \exp^{-(E_i/k_B T)} \tag{19}$$

Thus, an accurate evaluation of configuration entropy necessitates to span the whole system configuration space. AIMD [36] methods are perfectly designed to perform this task, but, once again, their computational cost is prohibitive for systems of realistic complexity. The *Cluster Variation Method* (CVM) [39] offers alternatives to the problem in providing analytical expressions of the internal energy, configurational entropy, and free energy of discrete lattice models in terms of cluster probability variables. The variable is here an occupation variable σ which defines one given distribution of Li and Li-vacancies over the n-sites of the cluster $\sigma = \{\sigma_1, \sigma_2, \ldots, \sigma_n\}$. This atomic feature makes the CVM coherent with first-principles calculations. Each cluster distribution is assigned to a given N-body interaction (pairs, triplets, quadruplets) for which an *Effective Cluster Interaction* (ECI) potential V can be fitted to accurate DFT calculations. The energy of the system is expanded in terms of the occupation variable and ECI, i.e., in terms of correlation cluster functions

$$E(\sigma) = E_0 + \sum_i V_i \sigma_i + \sum_{i,j \neq j} V_{i,j} \sigma_{i,j} + \ldots \tag{20}$$

For the expansion to be accurate, the sum must contain all the relevant N-body interactions required to describe short- to long-range inter-site correlations in the infinite lattice. The equilibrium state is then determined by a variational principle until full convergence of the ECIs. Once the ECIs are determined, the energy of any distribution can be easily extracted from Eq. (20). Very popular in material science, the CVM is being applied to battery materials with substitutional disorder not only to evaluate configuration entropy, [40, 41] but also to calculate phase diagrams, [42] to study ordering [43] or to model kinetic activation energies [41, 44]. Often coupled with Monte Carlo simulations, this method is regarded as the most reliable theoretical tool to deal with atomic correlations. However, since its accuracy relies on the truncation of the cluster expansion which is system dependent, the procedure to generate the clusters is not as straightforward as it appears to be and can still be very expensive. To reduce the computational cost and improve the predictive power of the cluster expansion, an elegant procedure based on a Bayesian approach has been proposed by Mueller and Ceder [45]. The main idea of this approach is to incorporate physical insights into the ECI fitting procedure to select the most pertinent cluster functions to be used in the training set.

Another way to reduce the computational cost of CVM is to use mean-field approaches. Among them, the Bethe–Peierls approximation [46] is considered as the lowest level of the CVM, as it is restricted to next-nearest neighbors correlations. The basic idea behind the Bethe–Peierls approximation is to compute the distribution probabilities of a given cluster using the next-nearest-neighbors interaction energies as obtained from first-principles DFT calculations and to use a self-consistent equation to further evaluate the inter-cluster energy as a function of the composition. Both the CVM and Bethe–Peierls mean-field approaches have been used to investigate the impact of configuration entropy on the finite-temperature phase diagram of Li-intercalated graphite [41, 47]. Finally, the easiest way to account for configuration entropy is to assume the distribution of Li^+/Li-vacancies over the crystal sites to be close to ideal solutions and to use the Stirling approximation to entropy

$$S = k_B(x \ln(x) - (1 - x) \ln(1 - x)) \qquad (21)$$

Because the Stirling approximation is asymptotically exact in the limit of weak and strong dilutions, it is generally sufficient to describe the single-phase domains that are expected to delimit a two-phase region. It is also considered as a reasonable approximation for evaluating configuration entropy in the limit of weak Li^+/Li-vacancies correlations.

In Li-battery materials, configuration entropy may affect the mechanism of the electrochemical reaction and is therefore a relevant quantity to be evaluated. First-principles calculations of Li-site energies and Li–Li correlations (in extended supercells) are valuable preliminary steps to check whether or not configuration entropy will contribute to the material free energy. In the limit of strong correlations, fully ordered phases are expected to be achieved upon Li-insertion so that configuration entropy can be neglected. In the limit of weak correlations, a rough estimate of $S_{conf}(T)$ is given by the Stirling approximation. In all other cases, a more accurate evaluation of configuration entropy is required.

3.3 Electrochemical Properties

Thermodynamic quantities
Once the overall electrochemical mechanism and the relevant intermediate Li-compositions are determined from the finite-temperature phase stability diagram, the equilibrium material properties such as the theoretical capacity and potential can be evaluated. When Li-insertion is predicted to follow a single-phase process, the Gibbs phase rule imposes that the material potential evolves as a function of the

reaction extent. Chemically speaking, this means that the redox entity is changing all along the electrochemical reaction. In contrast, when Li-insertion is predicted to follow a two-phase process, the electrode consists in a proportional mixture of the two end-members (the Li-rich and Li-poor phases) whose relative ratios are set by the reaction extent. In that case, the redox entity remains the same all along the electrochemical reaction and the Gibbs phase rule imposes that the material potential is constant in the composition range of the two-phase process. Following Eq. (4), the material potential with respect to the Li^+/Li^0 reference is linked to the molar reaction free energy $\Delta_r G(x)$:

$$
\begin{aligned}
V(x) &= -\frac{1}{F}\,\Delta_r G(x) = \sum_i \mu_i \nu_i \\
&= -\frac{1}{F}\left\{ \frac{\mu_{Li_{x_2}\mathcal{H}} - \mu_{Li_{x_1}\mathcal{H}} - (x_2 - x_1)\mu_{Li^0}}{(x_2 - x_1)} \right\}
\end{aligned}
\tag{22}
$$

where x_1 and x_2 refers to the Li-compositions considered in the reaction

$$
Li_{x_1}\mathcal{H} + (x_2 - x_1)Li \leftrightharpoons Li_{x_2}\mathcal{H} \tag{23}
$$

For a single-phase process, infinitesimal variations of the Li-content need obviously to be computed to predict the voltage profile of the $Li_x\mathcal{H}//Li^0$ battery. For a two-phase process, the battery voltage is set by the chemical potentials of the two Li-compositions delimiting the biphasic domain.

Within the DFT framework, the calculation of reaction enthalpies imposes that the energy (i.e., chemical potential) of the different phases involved in reaction (23) are computed using the same DFT functional (i.e., the same approximation for exchange and correlation energy). This condition can be restrictive when Li-insertion is accompanied with a change in the host material chemical bonds and electronic structure since in that case, different functionals or different parameter-izations of a given functional are in principle required to accurately describe the different phases involved in the electrochemical reaction. This restriction is par-ticularly severe for electrochemical reactions involving bond breaking or material decomposition such as conversion reactions, [48] but may also be significant in classical insertion reactions. In high-potential cathode materials such as 3d-transition metal oxides [49, 50], phosphates [51, 52], or sulfates [53, 54], the change in Li-composition is generally associated with a change in the transition metal formal oxidation state. Since metallic 3d-orbitals are spatially localized, the way they interact with the surrounding ligand orbitals may be strongly affected by the change in their occupation number. Hence, not only functionals going beyond the LDA and GGA approximations are required to properly describe the electronic structure of the two end-member compositions but also different parameterizations are necessary to account for the change in the transition metal oxidation state and 3d-orbital energies. For all these systems, reaction-free energies invariably contain a methodological error which is difficult to quantify, a priori. Alternatives exist to minimize this error but they are not transferable from one system to another. They are based on empirical corrections fitted on experimental data and applied to the free energy calculation [55].

In 1997, Ceder and coworkers were the first authors to use the Nernst relation to compute the average potential of electrode materials from first-principles [29]. In their work, they assumed a two-phase process between two perfectly ordered phases, e.g., a non-lithiated transition metal oxide (MO_y) and its fully lithiated homolog ($LiMO_y$) and neglected finite-temperature effects ($T = 0$ K calculations). Using the LDA approximation, they showed that the experimental voltages were systematically underestimated by the calculations with an error of ca. 0.3–0.4 V, which was first attributed to the poor description of the Li-metal cohesive energy [29]. The error was further reduced to a few tens or hundreds of millivolts with the use of more adapted DFT functionals such as the GGA+U [51] and HSE06 [52]. This is in perfect agreement with the physics included in these functionals to correct (at least partially) the SIE of conventional DFT (LDA or GGA).

> For materials showing limited structural reorganizations upon lithiation, $T = 0$ K first-principles DFT calculations generally predict electrochemical potentials within a few hundreds of millivolts. The remaining error arises not only from the unresolved methodological issues of the parameterized DFT functionals but also from any of the experimental setups that could affect the energy calculations while not included in the calculation, such as finite temperature effects. It is therefore important that these two errors are evaluated to reach meaningful and trustable material potentials.

To enable the discovery of new promising electrode materials, one strategy is to generate extended sets of potential candidates through "high-throughput" procedures and to take advantage of the stability of first-principles DFT calculations to compute their properties. The idea of building extended databases of calculated material properties and structural information has been initiated with the "Material project" of the MIT [33], and is being increasingly used by the Li-battery community [56–59]. Although this procedure appears much easier to conduct than experiments, it still requires a significant amount of work: both the Li-rich and the Li-poor compositions have to be computed within a reasonable numerical accuracy for their energy difference to be meaningful, and sometimes with finite-temperature effects when necessary. While this is certainly tractable for a quite large set of compounds [56, 57], it becomes limited or prohibitive when realistic materials including defects, disorder, or dopants must be studied [53, 60]. In these specific cases, an alternative strategy is to use the method discussed in Sect. 3.3 to rapidly assess the material potential. In case of strongly ionic systems, Eq. (8) restricts to electrostatic contributions that can be rapidly and accurately evaluated through the calculation of Madelung potentials at the sites where the electron and Li^+ are added:

$$V_{cell} \propto \frac{1}{F}(\mathcal{V}_{e^-} + \mathcal{V}_{Li}) \tag{24}$$

Applied to a wide variety of iron and cobalt-based materials (see Fig. 4), the method has proven to be accurate and efficient in reproducing the experimental data at a modest computational cost (several orders of magnitude lower than DFT calculations). In contrast to periodic DFT calculations that are unable to capture the effects of statistical cationic disorder on the material potential, this method allows the treatment of disordered materials through the use of fractional average charges in the calculation of the Madelung potentials. Its generalized expansion into meaningful and easily tunable quantities provides solid-state chemists new recipes for designing new electrode materials for Li-ion (or Na-ion) batteries. Coupled to crystallographic databases, such as the ICSD [61, 62] or COD [63, 64], it enables a powerful sampling of the most promising candidates for A-ion (A = Li, Na) battery materials through rapid assessment of their electrochemical potential.

Kinetic quantities Apart from the thermodynamic properties, kinetic quantities such as intrinsic Li-diffusivity can be investigated through first-principles DFT calculations. Because the timescale for Li-diffusion in bulk materials is usually several orders of magnitudes higher than the typical timescale of AIMD methods (~ 10 ps for a few hundreds of atoms per unit cell), alternative statistical approaches based on the Eyring equation [65] or on the Transition-State Theory [66] are used to assess the reaction rates of elementary diffusion events. These two methods were simultaneously developed to determine the minimum energy pathway (of the chemical reaction or diffusion event) between two local minima of the PES, and the associated transition state structure (TS).

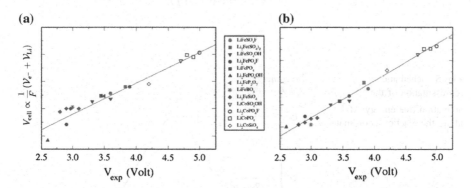

Fig. 4 Theoretical prediction of the electrochemical potential of various Fe- and Co-based polyanionic systems using Eq. (24). These materials exhibit high-potentials due to the strong electronic localization of the added electron on the transition metal center (RAC = Fe, Co). As a consequence, the Madelung potentials are here computed on the transition metal site ($\mathcal{V}_{e^-}$) and on the lithium site ($\mathcal{V}_{Li^+}$) of the Li-rich crystal structures using either formal charges (**a**) or Bader charges (**b**)

As shown in Fig. 5, the energy difference between the reactant structure (G^R) and the TS structure $(G^{\ddagger})$ gives the activation free energy $(\Delta G^{\ddagger})$ that is linked with the reaction rate (κ) through

$$
\kappa = \frac{k_B T}{h} \exp\left(-\frac{\Delta G^{\ddagger}}{k_B T}\right) = \frac{k_B T}{h} \exp\left(\frac{G^R - G^{\ddagger}}{k_B T}\right)
$$

$$
= \frac{k_B T}{h} \exp\left(\frac{\Delta S^{\ddagger}}{k_B}\right) \exp\left(-\frac{\Delta H^{\ddagger}}{k_B T}\right) \tag{25}
$$

$$
= v^{\ddagger}(T) \exp\left(-\frac{\Delta H^{\ddagger}}{k_B T}\right)
$$

Although formally equivalent to the empirical Arrhenius equation, this equation includes entropic and mechanistic considerations through the temperature-dependent pre-exponential factor $v^{\ddagger}(T)$ and the TS structure. $v^{\ddagger}$ is the so-called frequency factor. It depends on the partition functions of the reactant (Z^R) and TS $(Z^{\ddagger})$ structures which are primarily governed by the vibrations

$$
v^{\ddagger}(T) = \frac{k_B T}{h} \frac{Z^{\ddagger}}{Z^R} \sim \frac{k_B T}{h} \frac{Z^{\ddagger}_{\text{vib}}}{Z^R_{\text{vib}}} \tag{26}
$$

In condensed matter, Z_{vib} can be evaluated in the harmonic limit as

$$
Z_{\text{vib}} = \Pi_i \left(\frac{1}{1 - \exp\left(-\frac{h v_i}{k_B T}\right)}\right) \tag{27}
$$

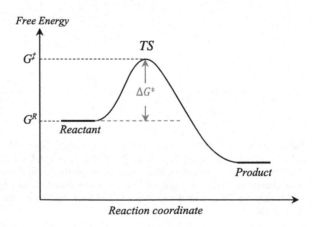

Fig. 5 Schematic representation of the activation free energy $\Delta G^{\ddagger}$ along the reaction coordinate

where v_i are the vibration frequencies. The diffusion coefficient (D) is then linked to the reaction rate (κ) through the hopping length (d)

$$D = d^2\kappa = d^2\left(\frac{k_BT}{h}\right)\frac{\Pi_{i=1}^{N}\left(1 - \exp\left(-\frac{hv_i^R}{k_BT}\right)\right)}{\Pi_{j=1}^{N-1}\left(1 - \exp\left(-\frac{hv_j^{\ddagger}}{k_BT}\right)\right)}\exp\left(\frac{H^R - H^{\ddagger}}{k_BT}\right) \qquad (28)$$

where N is the number of vibrational degrees of freedom of the reactant. Because the transition state structure is a saddle point of the system PES, it has one less vibrational degree of freedom than the reactant.

Practically, the easiest way to approach the reaction rate of Li-diffusion from first-principles is to use the *Nudged-Elastic Band* (NEB) model developed in the early 2000s by Henkelman et al. [67]. An initial guessed pathway is defined by a (linear) interpolation of the reactant and product optimized structures into M_i ($i = 1$ to n) regularly spaced structures named the "image" structures. The reactant and product relaxed structures are assigned to M_0 and M_{n+1} and are kept frozen during the procedure. The n other images are structurally relaxed under constraints imposed by adjacent images: each atom in image M_i ($i = 1$ to n) is virtually linked by harmonic forces (springs) with its associated atom in images M_{n+1} and M_{n-1}. The ensemble constituted by all images converges toward the minimal energy pathway between the reactant and product structures. In its initial version, the elastic band method was known to poorly describe the transition state structure and energy due to the tendency of images to slip down towards the reactant or product structures. To overcome this issue, more sophisticated constraints were developed to force the images to remain regularly spaced on the reaction pathway or to drive the highest energy image up to the saddle point TS structure [68]. Generally speaking, an NEB procedure is computationally expensive as it requires the optimization of $3Nn$ interdependent degrees of freedom (N = number of atoms per image). To reduce this computational cost, an alternative procedure is to perform "coarse" NEB without reaching full convergence and to re-optimize the as-obtained approximate TS structure. A vibration calculation of the final TS structure is recommended in this case to verify that this structure is a local extremum of the PES (one unique imaginary frequency).

Once the TS energy is known, the constant rate of each elementary event (or reaction) can be computed using the Arrhenius law and an appropriate parameterization of the pre-exponential factor. This generally gives reasonable estimates of the system kinetics as well as insights into the diffusion mechanisms for many applications. When more accurate reaction rates need to be extracted, the full vibrational structure of the system has to be computed (see Eq. (28)) to assess the pre-exponential factor and to include the finite-temperature dependence of enthalpy (H) as well as isotopic effects through the introduction of zero-point energy corrections (ZPE).

In battery materials, the calculation of activation barriers for various pathways gives insights into the Li-diffusion mechanisms. The constant rates and diffusion coefficients are determined within an error that depends on the TS structure energy through the choice of the DFT functional and on the way the pre-exponential factor is extracted. Since constant rates strongly depend on the Li-composition and local site occupancy, they need to be computed for different environments and Li-fractions. They can be interpolated using a CVM-like (Cluster Variation Method) approach such as that discussed in the previous section to build a composition and environment-dependent library of constant rates. The latter can further be used as inputs in kinetic Monte Carlo simulations [41, 69] to reach the macroscopic time scale by integrating the whole ensemble of environment-dependent diffusion rates under specific conditions of composition and external temperature.

4 Modelization of Interfaces

Nowadays, first-principles electronic structure calculations can be helpful in gaining information about the structure and energy of solid/solid and solid/liquid interfaces. In Li-ion batteries, the electrochemical reactions taking place at both electrodes can be significantly altered by the electrode morphology, active material particles sizes, and electrolyte composition. Not only the kinetics of the electrochemical reactions but also the thermodynamics is modified by the particles size [70]. More specifically, side reactions occurring at the electrode/electrolyte interfaces or Li-dendrite formation are responsible for degradation or ageing phenomena which, in turns, affect the battery performance [71–73]. Reaching an atomistic description of these mechanisms is therefore central to the ongoing research in this field. Significant advances from the computational side have recently been reported together with the emergence of multiscale models. While it is not the purpose of this chapter to give an exhaustive review of the computational techniques addressing surface and interface science, we believe it is important to stress several differences between bulk thermodynamics and surface thermodynamics. Interested readers can refer to the pioneer work of Schmickler on surface electrochemistry [74].

4.1 Surface/Interface Thermodynamics

As shown in the previous section, the free energy of a bulk material is primarily governed by its chemical composition and crystal structure and to a lower extent by finite-temperature effects (see Eq. (2)). The electrochemical properties of bulk materials are therefore easy to address through the computation of composition-dependent bulk free energies. In the case of surfaces, the surface charge (Q_s) is an additional degree of freedom of the surface free energy

$$G(\bar{x}, T, Q_s) = n_b \mu_{\mathrm{Li}_{x_b} \mathcal{H}}(x_b, T) + n_s \mu_{\mathrm{Li}_{x_s} \mathcal{H}}(x_s, T, Q_s)$$
$$+ (n_{\mathrm{Li}} - n_T.\bar{x}) \mu_{\mathrm{Li}} \tag{29}$$

where $n_T = n_b + n_s$ is the number of mole of reacting Li in bulk and surface (in mole) and x_b and x_s their molar fraction. The reaction extent is here set by $n_T \cdot \bar{x}$ where $\bar{x} = (n_b \cdot x_b + n_s \cdot x_s)/(n_b + n_s)$ is the average Li composition of the electrode. At equilibrium, the Li chemical potential is equal in the bulk phase $\left(\mu_{\mathrm{Li}}^{\mathrm{Li}_{x_s} \mathcal{H}} \right)$ and in the surface phase $\left(\mu_{\mathrm{Li}}^{\mathrm{Li}_{x_b} \mathcal{H}} \right)$. Using Eq. (4) we can then write

$$\frac{\partial \mu_{\mathrm{Li}_{x_s} \mathcal{H}}(x_s, T, Q_s)}{\partial x_s} - \mu_{\mathrm{Li}} = \frac{\partial \mu_{\mathrm{Li}_{x_b} \mathcal{H}}(x_b, T, Q_s = 0)}{\partial x_b} - \mu_{\mathrm{Li}} = -FV_{\mathrm{cell}}(\bar{x}) \tag{30}$$

to link the Li chemical potential to the battery voltage at composition $\bar{x}$. Similar to the bulk case, $V_{\mathrm{cell}}(\bar{x})$ refers to the difference in the two electrode potentials, that is, $V_{\mathrm{cell}}(\bar{x}) = V_s - V_{\mathrm{Li}^+/\mathrm{Li}^0}$ where V_s is the surface potential. In first approximation the surface potential is linked to the surface charge by the linear relation

$$(V_s - V_s^0) = Q_s/C_s \tag{31}$$

where V_s^0 is the zero-charge surface potential and C_s the surface differential capacitance. The electrical work to charge the surface from neutrality to Q_s is then

$$W_e \sim Q_s \cdot V_0 + \frac{Q_s^2}{2C_s} \tag{32}$$

This highlights an important difference between bulk and surfaces electrochemistry. At equilibrium, the Li-poor $(\mathcal{H}_s)$ and Li-rich $(\mathrm{Li}\mathcal{H}_s)$ surfaces must have the same potential, while they have no reason to be equivalently charged. As schematically shown in Fig. 6, the reaction free energy must now include the modification of the electrical work to bring the two surfaces to the same potential. The elementary surface reaction thus corresponds to

$$\mathcal{H}_s^{Q_p} + \mathrm{Li}^+ + \left(1 + \frac{(Q_p - Q_r)}{F} \right) e^- \leftrightarrows \mathrm{Li}\mathcal{H}_s^{Q_r} \tag{33}$$

where Q_p and Q_r are the Li-poor and Li-rich surface charges, respectively.

This expresses the surface polarization induced by the elementary reaction which here corresponds to Li-adsorption on the surface. This polarization is negligible only when the Li-rich and Li-poor surfaces have similar zero-charge potentials.

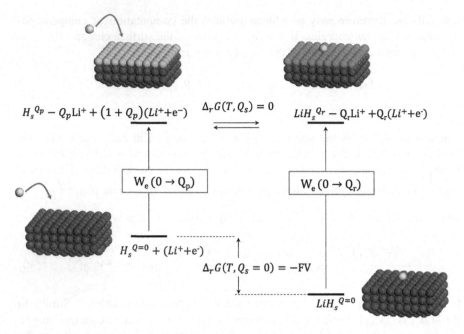

$$H_s^{Q_p} - Q_p Li^+ + (1 + Q_p)(Li^+ + e^-) \quad \Delta_r G(T, Q_s) = 0 \quad LiH_s^{Q_r} - Q_r Li^+ + Q_r(Li^+ + e^-)$$

$$W_e(0 \to Q_p) \qquad\qquad W_e(0 \to Q_r)$$

$$H_s^{Q=0} + (Li^+ + e^-)$$

$$\Delta_r G(T, Q_s = 0) = -FV$$

$$LiH_s^{Q=0}$$

Fig. 6 Illustration of reaction (33) where W_e correspond to the electrical works associated with the charging of $\mathcal{H}_s$ and $Li\mathcal{H}_s$ from zero to Q_p and Q_r respectively. The reaction free energy is indicated for the zero-charged surfaces $\Delta_r G(T, Q_s = 0)$ and for the charged surfaces. $\Delta_r G(T, Q_s)$

In all other cases, the reaction free energy is linked to the zero-charge reaction free energy through

$$\Delta_r G(\bar{x}, T, Q_s) = \Delta_r G(\bar{x}, T, Q_s = 0) + \delta W_e + F V_{\text{cell}}(\bar{x}) = 0 \qquad (34)$$

where

$$\delta W_e = W_e(0 \to Q_r) - W_e(0 \to Q_p)$$
$$\sim \frac{1}{F}\left(Q_r(V_{Li\mathcal{H}_s}^0 - V_{Li^+/Li^0}) - Q_p(V_{\mathcal{H}_s}^0 - V_{Li^+/Li^0}) \right) \qquad (35)$$

is the difference in the electrical work to reach the equilibrium potential $(V_{\text{cell}}(\bar{x}))$ and the associated surface charges Q_p and Q_r. The cell voltage is then

$$V_{\text{cell}}(\bar{x}, T) = -\frac{1}{F}\left(\Delta_r G(\bar{x}, T, Q_s = 0) + \delta W_e \right) \qquad (36)$$

4.2 First-Principles Approach to Charged Surfaces

In practice, several approaches have been proposed to address surface electrochemistry through first-principles DFT calculations [75–77]. The first one has been proposed by Nørskov et al. [75] and does not include the variation in the surface charge density due to the variation of the electrochemical potential. The system energetics then restricts to the zero-charge energetics

$$V_{cell}(\bar{x}, T) \sim -\frac{1}{F}\Delta_r G(\bar{x}, T, Q_s = 0) \tag{37}$$

This approximation has proven to yield reliable results in fuel cells to describe the oxygen reduction reaction (ORR) which involves intermediate surfaces with close zero-charge potentials [75]. More recent applications have shown that while the surface polarization is important in an absolute sense, the changes in reaction energies are not as sensitive provided that the reactant and products have similar directional dipoles. In Li-ion batteries, this approximation may fail each time a significant modification of the surface polarization or orientation of the surface dipole is expected upon lithiation. In this case, it is crucial to compute charged surfaces to account for the electric work contribution to the free energy. This can be done following the general procedure proposed by Filhol and coworkers [77–79]. In this method, a bias is introduced by adding (withdrawing) electron to (from) the unit cell that are compensated by a homogeneous background charge to ensure cell electroneutrality. Since the homogeneous background contributes to the total energy through nonphysical interactions with the slab, an energy correction to the system energy must be added. If N_e is the number of excess electrons (holes) and $N_{bg} \cdot e = -N_e \cdot e$ the associated background charge, the total energy of the surface, as obtained from a DFT calculation is written

$$dE_{DFT} = \mu_e dN_e + \mu_{bg} dN_{bg} = (\mu_e - \mu_{bg})dN_e \tag{38}$$

The chemical potential of the homogeneous background charge is given by its electrostatic interaction with the rest of the system

$$\mu_{bg} = \frac{1}{\Omega}\int_\Omega V(\vec{r}, N_e)d\vec{r} = \bar{V}(N_e) \tag{39}$$

where Ω is the unit cell volume, $V(\vec{r}, N_e)$ the electrostatic potential (with respect to the reference potential[3]) and $\bar{V}(N_e)$ the average potential in the unit cell for N_e excess electrons.

---

[3]In surface calculations, boundary conditions are applied to a unit cell consisting in a material slab characterized by its (hkl) Miller indices and a vacuum layer. The spatial reference for the potential is set by the position in the vacuum layer where the electrostatic potential is a local extremum.

The homogeneous background can be seen as a 0-order model for the Helmholtz double layer in the solvent part. It induces another nonphysical feature in the metal slab since a fraction of the added electrons is used to screen the background in the slab, at equilibrium. As a consequence, the remaining active electrons for the electrochemical reaction (N_e^s) are estimated as

$$N_e^s \sim \frac{z_0}{c} N_e \tag{40}$$

where c is the unit-cell parameter in the z-direction and z_0 is the interslab distance.[4] The energy of the charged surface then refers to the zero-charged surface energy through

$$E_{el}^s(N_e) = E_{DFT}(0) + \frac{z_0}{c}\left(E_{DFT}(N_e) - E_{DFT}(0) + e\int_0^{N_e} \bar{V}(N_e)dN_e\right) \tag{41}$$

This correction checks the fundamental thermodynamic relation between the surface energy, surface charge, and electrochemical potential μ_e

$$\left(\frac{\partial E_{el}^s}{\partial N_e^s}\right) = \left(\frac{\partial E_{el}^s}{\partial N_e}\right) \times \left(\frac{\partial N_e}{\partial N_e^s}\right)$$

$$= \frac{z_0}{c}\frac{\partial}{\partial N_e}\left(E_{DFT}(N_e) + e\int_0^{N_e} \bar{V}(N_e)dN_e\right)\frac{c}{z_0} \tag{42}$$

$$= \mu_e$$

The approximation of Eq. (42) leads to nearly identical numerical results than other approaches that extract this number more precisely from the surface electric field or energy derivative. In the case of solid/vacuum interfaces, it can be seen as an extension of surface modeling approaches earlier developed to study catalytic effects. The method has been shown to reach a spectroscopic accuracy in reproducing and understanding complex electrochemical effects such as the Stark effects, that is, the modification of vibrational frequencies of surface-adsorbed species with the applied bias [80]. It is also a powerful tool to model complex phase mixing

---

(Footnote 3 continued)

Surfaces are generally built in such a way that the two surfaces of the active slab do not interact and are symmetrically related. This way, the middle of the interslab distance corresponds to the reference potential position.

[4]The interslab distance must account for the slab extension and is therefore calculated as the difference in the z-coordinate of the atoms lying on either sides of the layer minus their atomic radii.

occurring for water deposits on metal surfaces [78, 81, 82]. This energy correction can be straightforwardly implemented in DFT codes and applied to more complex interfaces such as solid/solid [83] and solid/liquid [84] interfaces occurring in Li-ion batteries.

4.3 Application to Solid/Liquid Interfaces

In Li-ion batteries, the current generation of electrolytes consists in an Li-salt, generally $LiPF_6$ solvated in alkyl-carbonates solvent molecules. The total capacitance of the solid/electrolyte interface (C) can be decomposed into the intrinsic surface capacitance (C_s) and the double layer capacitance (C_{DL}) through $1/C = 1/C_s + 1/C_{DL}$. Therefore, it is not an intrinsic surface property but depends on the electrolyte concentration through the double layer composition and structuration of the double layer. In the limit of concentrated electrolyte (e.g., 1 mol L^{-1}), the contribution of the double layer capacitance can be neglected so that $C \sim C_s$. One drawback of vacuum approaches to model solid/liquid interfaces is that the double layer capacitance of vacuum is in the same range of C_s and can no longer be neglected. It is at least one order of magnitude smaller than that of current electrolytes. The difference between a vacuum layer and a solvent lies in their dielectric constants that differ by nearly two orders of magnitude (typically 1 for vacuum and 90 for the current generation of solvents).

Another issue related to vacuum approaches is that the potential of electrochemical reactions is not always reliable when the process involves solvated ionic species. As an example, the electrochemical potential of the Li^+/Li^0 redox couple is found to be 2 V with respect to the normal hydrogen electrode (NHE) which would imply that Li^+ is a strong oxidant! It is therefore mandatory to account for the solvent effects to recover both reliable potentials and double layer capacitance. Practically, this can be done by adding many layers of solvent over the surface which is computationally very demanding and does not allow a description of the solvent in its full structural and time-dependent complexity [85, 86]. Another approach is to use an implicit continuum representation of the solvent molecules as routinely done in molecular calculations, [87, 88] and in some condensed matter approaches [89, 90]. While the applicability of the implicit solvent models is well defined for neutral molecules and surfaces, its pertinence to address charged surface properties has only recently been validated in the test case of Li-metal/electrolyte electrochemical interface [84]. For this complex interface, the cavity size parameter of the PCM has been shown to strongly influence the surface capacitance and needs to be carefully checked. While PCM is able to properly account for the electrostatic properties of the solvent, it misses the bonding dimension. Given the $Li^+ + e^- \rightarrow Li$ reaction occurring at the Li-metal/electrolyte interface, the PCM approach leads to an equilibrium potential of -1.6 V versus NHE ($+1.44$ vs. Li^+/Li^0). While this constitutes an important improvement compared to vacuum calculations ($+2$ V vs. NHE) it is still

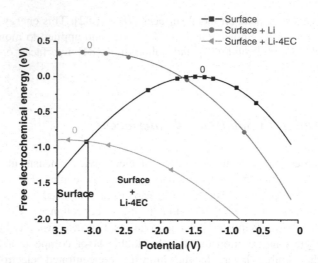

Fig. 7 Free electrochemical energy (eV) as a function of the applied potential (with respect to NHE) for the Li-metal surface in contact with a purely implicit PCM solvent (surface), an Li$^+$ immersed in PCM without (Li-surface) and with its explicit first solvation shell of 4 ethylene carbonate molecules (Li-4EC). For each potential, the surface with the lowest free energy is the most stable

far away from the reference and shows that at least the first solvation shell of Li$^+$ needs to be explicitly added to the system. As shown in Fig. 7, the explicit addition of four solvent molecules, e.g., ethylene-carbonate (EC) around the Li$^+$ leads to -3.1 V versus NHE which is only 60 mV below the experimental value. For this textbook study, the surface polarization mentioned above is computed as

$$S^{0.85-} + \text{Li(EC)}_4^{0.95+} + 1.1e^- \leftrightarrows \text{LiS}^- + 4\text{EC} \tag{43}$$

where LiS$^-$ corresponds to the charged surface with Li inserted in the bulk phase. Using this electrochemical approach together with a semi-explicit solvent approach allows not only recovering equilibrium properties but also computing potential-dependent energy barriers associated with the elementary steps of the electro-chemical reaction [91]. In this way, the microscopic potential-dependent kinetics of the system can be integrated in Monte Carlo simulations to reach macroscopic timescales in a multiscale procedure.

4.4 Application to Solid/Solid Interfaces

In contrast to solid/liquid interfaces, solid/solid interfaces may exhibit large mechanical stress due to crystallographic mismatches between the two interacting surfaces. For nanosized systems, the interplay between mechanical stress, interface energy, and local electric field is crucial and may affect the system thermodynamics and kinetics. From a general point of view, a solid/solid interface is characterized by three leading interdependent descriptors: the chemical, the mechanical, and the electrical descriptors.

The former arises from chemical bonding between the two side-members of the interface that primarily controls adhesion properties. The interfacial relaxation is a direct consequence of the different atomic environments compared to isolated bulk and might lead to large modifications of bond strength and/or connectivity at the interface. The energy associated with this chemical descriptor is the interface energy γ, generally expressed in J/m^2 which tunes the adhesion of the interacting phases, and therefore dictates the electrode morphology in multiphase systems. In the simple case of two interacting phases, A and B, immersed into an electrolyte E, different situations may occur depending on the relative interface energies (see Fig. 8): (a) If $\gamma_{A/E}$ and $\gamma_{B/E}$ are lower than $\gamma_{A/B}$, A and B will separate to maximize the solid/electrolyte contact areas; (b) if $\gamma_{A/B}$ is lower than $\gamma_{A/E}$ and $\gamma_{B/E}$, the two phases will merge or aggregate to minimize the contact area with the solvent; (c) if $\gamma_{A/B}$ is negative, the two phases will mix to maximize their contact area, and the one exhibiting the highest interface energy with the solvent will be embedded into the other. Note that when the energy of all possible interfaces is known, the equilibrium shape/morphology of the whole system can be determined using a Wulff-like approach.

The second descriptor characterizing an interface is the mechanical stress induced by the mismatch between A and B (see Fig. 8). Due to this mismatch, an extended strain may propagate to a large volume from the interface toward the two interacting phases (for sake of clarity, the strain is shown only for B in Fig. 8). Usually, lattice mismatches smaller than a few percent lead to epitaxial interface structures. For higher mismatches, important macroscopic elastic stress arises which can be partially released by the creation of dislocations. These dislocations occur when the stress energy release is larger than the energy cost of the dislocation core creation (bond breaking). In the case of small particles, the stress accumulated in this volume remains small enough to prevent dislocations. In this case, the mechanical stress contributes to the total free energy of the system, therefore shifting the electrochemical potential from the extended bulk value.

Finally, the last descriptor of an interface arises from the electric contact between the two interacting phases. As for any heterojunction, the equilibrium between two interacting phases results in a residual interface dipole $\vec{p}$ (see Fig. 8) which has two origins: the first one is local and results from the polarity of the interfacial bonding; the other is nonlocal and originates from the difference in Fermi levels of the two isolated phases. When an interface is created, an interphase charge transfer occurs to equilibrate the Fermi levels, thus achieving a homogeneous electrochemical potential over the whole system. This leads to an electrode polarization illustrated by positive and negative charges on Fig. 8. Because these charges remain localized near the interface, the resulting dipole $\vec{p}$ may generate strong local electric fields that can be responsible for an enhanced electrochemical reactivity.

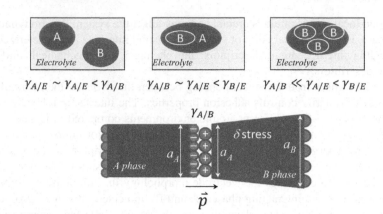

$\gamma_{A/E} \sim \gamma_{A/E} < \gamma_{A/B}$ $\quad$ $\gamma_{A/B} \sim \gamma_{A/E} < \gamma_{B/E}$ $\quad$ $\gamma_{A/B} \ll \gamma_{A/E} < \gamma_{B/E}$

Fig. 8 Illustration of the electrode morphology depending on the relative energies of the solid/solid interfaces (A/B) and the solid/electrolyte interfaces (A/E, B/E) for a simple case of two phases immersed into a solvent. Schematic representation of the elastic strain of an interface and the resulting electric dipole $\vec{p}$

Accounting for these descriptors from first principles is very challenging, as witnessed by the rather small number of studies reported so far in the field of Li-ion batteries. Moreover, these interdependent descriptors are difficult to extract independently. Superlattice approaches have recently been proposed to investigate complex multiphased electrodes such as conversion reactions [83]. These reactions are prototypes for studying solid/solid interface electrochemistry since the extent of the reaction is dictated by front phase migrations between the three involved phases, i.e., the starting MX and the composite M^0/LiX. In the superlattice multi-interface approach proposed by Dalverny et al. [83], the *chemical, mechanical,* and *electrical* descriptors were shown to be easily extracted from first-principles DFT calculations and used as tools for understanding the electrochemical mechanisms. Thanks to a formal decomposition of the interfaces energies into one *chemical* and one *mechanical* contribution, the energetical hierarchy of the various interfaces was used to predict the electrode morphology (see Fig. 9). The cell voltage was then extrapolated from the elastic stress energy induced by the interfacial strain. External oxidizing or reducing conditions were then applied to the multi-interface superlattice in order to follow the structural and electronic response of the whole system upon charge/discharge. This *electrical* descriptor is central in surface electrochemistry [92] and was shown in the specific case of conversion reactions to exhibit an asymmetric behavior upon charge and discharge, therefore rationalizing the origin of the voltage hysteresis experimentally observed in these systems [48, 83, 93, 94]. The voltage hysteresis was further quantified through a mechanistic study of the surface elementary reaction steps, using the Nørskov approximation discussed in Sect. 4.2. All the parameters extracted from this methodology can be used as input parameters in multiscale approaches to reach larger-scale phenomena such as aging or degradation mechanisms.

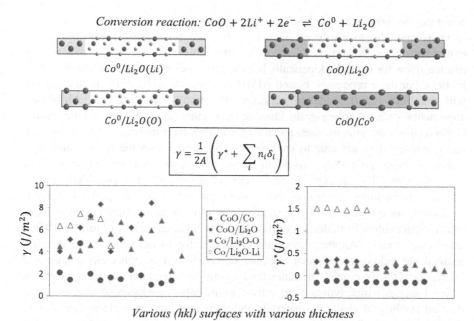

Various (hkl) surfaces with various thickness

Fig. 9 Representation of the different interfaces occurring during conversion of the CoO electrode into Co^0 and Li_2O. The interfaces are built in such a way that the common interface area (A) minimizes the elastic stress on the two end-member phases. The lithium oxide surfaces can be terminated by either a lithium plane ($Li_2O(Li)$) or an oxygen plane ($Li_2O(O)$). The interface energy γ is decomposed into one chemical contribution (γ^*) which corresponds to the interface adhesion and one mechanical contribution (δ) associated with the strain imposed to the end-member phases to remove the mismatch. While the interface energies (γ)—as computed from DFT calculations on various (hkl) planes for CoO, Co^0 and Li_2O and various layer thickness—are spread over a large range of energy with no energetical hierarchy, the adhesion energy γ^* clearly recovers a chemically intuitive hierarchy: the oxide/oxide interfaces make strongly ionic Li–O and Co–O bonds and are the most stable. Then, the $CoO/Li_2O(O)$ and Co^0/CoO have equivalent energies as they both form equivalent Co–O bonds. The less stable interfaces are those involving Co–Li bonds which are known to be unstable (no stable Li/M(3d) alloy reported in the literature)

5 Perspectives

This chapter aims at giving an overview of the state-of-the-art first-principles methodologies developed so far to address the complexity of electrochemical systems. Bulk versus interface electrochemistry has been tackled to highlight their similarities and differences from a physical and technical perspective. The link between basic thermodynamic and first-principles approach to condensed mater has been given. It was our choice not to give an extensive review of the theoretical studies reported on specific electrochemical systems but to focus on general approaches to these systems along with their related numerical and methodological issues. AIMD methods were not discussed in this chapter but should become central to the ongoing research in this field in the next decades with the increase of computer performance. They will enable reaching more accurate descriptions of solvents and electrode/solvent interfaces that are untractable today by purely explicit methods. Nevertheless, the timescale achievable by AIMD will remain much too low for studying kinetically limited processes associated with high kinetic barriers. For these processes, biased AIMD can provide interesting alternatives but will miss the interdependence between all competitive processes. Multiscale approaches such as coarse-grain classical molecular dynamics or kinetic Monte Carlo methods are able to reach much larger length and timescales, but the physics and chemistry they are able to capture strongly depends on the quality and relevance of input parameters. Most of these parameters can be extracted from first-principles DFT calculations on systems of reasonable size, provided that the relevant interactions governing the system properties are properly identified. A challenging question for the next decade is how to automatically extract these relevant parameters to build appropriate big-data libraries which, in turn, will feed multiscale models. Another important challenge for theoreticians is to develop adapted methodologies to reach a proper description of complex electrochemical interfaces. This is crucial to understand aging and degradation mechanisms in Li-ion batteries which typically arise from multiscaled interfacial phenomena. The thermal stability of electrode materials and their reactions with electrolyte should also be systematically considered as a possible cause for battery degradation. In this regard, one strategy is to build models to identify the leading parameters that govern the material reactivity as a function of operating conditions (potential, particle-size, electrolyte composition) and to further evaluate the thermodynamic and kinetic associated parameters through high-throughput procedures for large sets of candidates. The development of chemical, mechanical, and electrical descriptors appears as a top priority to enable a rational design of the next generation of electrode materials for Li/Na batteries.

References

1. Plett GL (2004)Extended Kalman filtering for battery management systems of LiPB-based HEV battery packs Part 2. Modeling and identification. J Power Sources 134(2):262–276
2. Doyle M, Newman J (1995) Modeling the performance of rechargeable lithium-based cells: design correlations for limiting cases. J Power Sources 54(1):46–51
3. Darling R, Newman J (1998) Modeling Side Reactions in Composite LiyMn$_2$O$_4$ Electrodes. J Electrochem Soc 145(3):990–998
4. Ramadass P, Haran B, Gomadam PM, White R, Popov BN (2004) Development of first principles capacity fade model for Li-Ion cells. J Electrochem Soc 151(2):A196–A203
5. Safari M, Morcrette M, Teyssot A, Delacourt C (2009) Multimodal physics-based aging model for life prediction of Li-Ion batteries. J Electrochem Soc 156(3):A145–A153
6. Safari M, Delacourt C (2011) Mathematical Modeling of Lithium Iron Phosphate Electrode: Galvanostatic Charge/Discharge and Path Dependence. J Electrochem Soc 158(2):A63–A174
7. Young WM, Elcock EW (1966) Monte Carlo studies of vacancy migration in binary ordered alloys: IProc Phys Soc 89:735–746

8. Fowler R, Guggenheim EA (1939) Statistical thermodynamics. Cambridge University Press, Cambridge, p 7
9. Lebon G, Jou D, Casas-Vázquez J (2008) Understanding non-equilibrium thermodynamics: foundations, applications, frontiers. Springer, Berlin. ISBN: 978-3-540-74252-4
10. Karma A, Rappel W-J (1998) Quantitative phase-field modeling of dendritic growth in two and three dimensions. Phys Rev E 57(4):4323–4349
11. Franco AA, Schott P, Jallut C, Maschke B (2007) A multi-scale dynamic mechanistic model for the transient analysis of PEFCs. Fuel Cells 7(2):99–117
12. Bazant M (2013) Theory of chemical kinetics and charge transfer based on nonequilibrium thermodynamics. Acc Chem Res 46(5):1144–1160
13. Franco AA (2013) Multiscale modelling and numerical simulation of rechargeable lithium ion batteries: concepts, methods and challenges. RSC Adv 3:13027–13058
14. Padhi AK, Nanjundaswamy KS, Masquelier C, Okada S, Goodenough JB (1997) Effect of structure on the Fe3+/Fe2+ Redox Couple in Iron Phosphates. J Electrochem Soc 144:1609–1613
15. Saubanère M, Ben Yahia M, Lebègue S, Doublet M-L (2014) An intuitive and efficient method for cell voltage prediction of lithium and sodium-ion batteries. Nat Commun 5:5559
16. Hohenberg P, Kohn W (1964) Inhomogeneous electron gas. Phys Rev 136(3):B864–871
17. Kohn W, Sham LJ (1965) Self-Consistent equations including exchange and correlation Effects. Phys Rev 140(4):A1133–A1138
18. von Barth U, Hedin L (1972) A local exchange-correlation potential for the spin polarized case. J Phys C: Solid State Phys 5(13):1629–1642
19. Wang Y, Perdew JP (1991) Correlation hole of the spin-polarized electron gas, with exact small-wave-vector and high-density scaling. Phys Rev B 44(24):13298–13307
20. Perdew JP, Wang Y (1992) Accurate and simple analytic representation of the electron-gas correlation energy. Phys Rev B 45(23):13244–13249
21. Becke AD (1993) Density functional thermochemistry III. The role of exact exchange. J Chem Phys 98(7):5648–5652
22. Perdew JP, Ernzerhof M, Burke K (1996) Rationale for mixing exact exchange with density functional approximations. J Chem Phys 105(22):9982–9985
23. Adamo C, Barone V (1999) Toward reliable density functional methods without adjustable parameters: The PBE0 model. J Chem Phys 110(13):6158–6170
24. Heyd J, Scuseria GE, Ernzerhof M (2003) Hybrid functionals based on a screened coulomb potential. J Chem Phys 118(18):8207–8215
25. Anisimov VI, Zaanen J, Andersen OK (1991) Band theory and Mott insulators: Hubbard U instead of Stoner I. Phys Rev B 44(3):943–954
26. Liechtenstein AI, Anisimov VI, Zaanen J (1995) Density-functional theory and strong interactions: orbital ordering in Mott-Hubbard insulator. Phys Rev B 52(8):R5467–R5470
27. Dudarev SL, Botton GA, Savrasov SY, Humphreys CJ, Sutton AP (1998) Electron-energy-loss spectra and the structural stability of nickel oxide: An LSDA+U study. Phys Rev B 57(3):1505–1509
28. Kulik HJ, Cococcioni M, Scherlis DA, Marzari N (2006) Density functional theory in transition-metal chemistry: a self-consistent Hubbard U approach. Phys Rev Lett 97 (10):103001-1-4
29. Aydinol MK, Kohan AF, Ceder G, Cho K, Joannopoulos J (1997) Ab initio study of lithium intercalation in metal oxides and metal dichalcogenides. Phys Rev B 56(3):1354–1365
30. Feynman RP (1939) Forces in Molecules. Phys Rev 56(4):340–343
31. Oganov AR, Glass CW (2006) Crystal structure prediction using ab initio evolutionary techniques: Principles and applications. J Chem Phys 124(24):244704-1–15
32. Ceder G, Morgan D, Fischer C, Tibbetts K, Curtarolo S (2006) Data-Mining-Driven quantum mechanics for the prediction of structure. MRS Bull 31:981–985
33. See the Material Project website: https://www.materialsproject.org/
34. Kirkpatrick S, Gelatt CD, Vecchi MP (1983) Optimization by simulated annealing. Science 220(4598):671–680

35. Wales D, Doye J (1997) Global Optimization by basin-hopping and the lowest energy structures of lennard-jones clusters containing up to 110 Atoms. J Phys Chem A 101 (28):5111–5118
36. Grotendorst J (ed) (2000) Modern methods and algorithms of quantum chemistry. John von Neumann Institute for Computing, Jülich, NIC Series, Vol 1, pp 301–449. ISBN 3-00-005618-1
37. Gonze X, Lee C (1997) Dynamical matrices, Born effective charges, dielectric permittivity tensors, and interatomic force constants from density-functional perturbation theory. Phys Rev B 55(16):10355–10368
38. Ben Yahia M, Lemoigno F, Beuvier T, Filhol JS, Richard-Plouet M, Brohan L, Doublet ML (2009) Updated references for the structural, electronic, and vibrational properties of $TiO_2(B)$ bulk using first-principles density functional theory calculations. J Chem Phys 130 (20):204501-1-11
39. Kikuchi R (1951) A Theory of Cooperative Phenomena. Phys Rev 81(6):988–1003
40. Sanchez JM, Ducastelle F, Gratias D (1984) Generalized cluster description of multicomponent systems Physica 128:334–350
41. Persson K, Hinuma Y, Meng YS, Van der Ven A, Ceder G (2010) Thermodynamic and kinetic properties of the Li-graphite system from first-principles calculations. Phys Rev B 82 (12):125416-1-9
42. Van der Ven A, Aydinol MK, Ceder G, Kresse G, Hafner J (1998) First-principles investigation of phase stability in Li_xCoO_2. Phys Rev B 58(6):2975–2987
43. Van der Ven A, Aydinol MK, Ceder G (1998) First Principles evidence for stage ordering in Li_xCoO_2. J Electrochem Soc 145(6):2149–2155
44. Van der Ven A, Ceder G, Asta M, Tepesch PD (2001) First-principles theory of ionic diffusion with nondilute carriers. Phys Rev B 64(18):184307-1-17
45. Mueller T, Ceder G (2009) Bayesian approach to cluster expansions. Phys Rev B 80 (2):024103-1-13
46. Bethe HA (1935) Statistical theory of superlattices. Proc Roy Soc London A 150:552–575
47. Filhol J-S, Combelles C, Yazami R, Doublet M-L (2008) Phase diagrams for systems with low free energy variation: A coupled theory/experiments method applied to Li-graphite. J Phys Chem C 112(10):3982–3988
48. Cabana J, Monconduit L, Larcher D, Palacin MR (2010) Beyond Intercalation-Based Li-Ion Batteries: The State of the art and challenges of electrode materials reacting through conversion reactions. Adv Mater 22:E170–E192
49. Wang L, Maxisch T, Ceder G (2006) Oxidation energies of transition metal oxides within the GGA+U framework. Phys Rev B 73(19):195107-1-6
50. Chevrier VL, Ong SP, Armiento R, Chan MKY, Ceder G (2010) Hybrid density functional calculations of redox potentials and formation energies of transition metal compounds. Phys Rev B 82(7):075122-1-11
51. Zhou F, Cococcioni M, Marianetti CA, Morgan D, Ceder G (2004) First-principles prediction of redox potentials in transition-metal compounds with LDA+U. Phys Rev B 70(23):235121-1–8
52. Hautier G, Jain A, Ong SP, Kang B, Moore C, Doe R, Ceder G (2011) Phosphates as Lithium-Ion Battery Cathodes: An evaluation based on high-throughput ab Initio calculations. Chem Mater 23:3495–3508
53. Ben Yahia M, Lemoigno F, Rousse G, Boucher F, Tarascon JM, Doublet ML (2012) Origin of the 3.6 V to 3.9 V voltage increase in the $LiFeSO_4F$ cathodes for Li-ion batteries. Energy Environ Sci 5:9584–9594
54. Frayret C, Villesuzanne A, Spaldin N, Bousquet E, Chotard J-N, Recham N, Tarascon J-M (2010) $LiMSO_4F$ (M = Fe, Co and Ni): promising new positive electrode materials through the DFT microscope. Phys Chem Chem Phys 12:15512–15522
55. Jain A, Hautier G, Ping Ong S, Moore CJ, Fischer CC, Persson KA, Ceder G (2011) Formation enthalpies by mixing GGA and GGA + U calculations. Phys Rev B 84 (4):045115-1-10

56. Hautier G, Fischer C, Ehrlacher V, Jain A, Ceder G (2011) Data Mined ionic substitutions for the discovery of new compounds. Inorg Chem 50(2):656–663
57. Mueller T, Hautier G, Jain A, Ceder G (2011) Evaluation of tavorite-structured cathode materials for lithium-Ion batteries using high-throughput computing Chem Mater 23 (17):3854–3862
58. Jain A, Hautier G, Moore CJ, Ping Ong S, Fischer CC, Mueller T, Persson KA, Ceder G (2011) A high-throughput infrastructure for density functional theory calculations Comp Mater Sci 50:2295–2310
59. Ping Ong P, Richards WD, Jain A, Hautier G, Kocher M, Cholia S, Gunter D, Chevrier V, Ceder G (2013) Python Materials Genomics (pymatgen): A robust, open-source python library for materials analysis. Comp Mater Sci 68:314–319
60. Van der Ven A, Ceder G (2005) Vacancies in ordered and disordered binary alloys treated with the cluster expansion. Phys Rev B 71(5):054102-1-7
61. Bergerhoff G, Hundt R, Sievers R, Brown ID (1983) The inorganic crystal structure data base. J Chem Inf Comput Sci 23:66–69
62. Belsky A, Hellenbrandt M, Karen VL, Luksch P (2002) New developments in the inorganic crystal structure database (ICSD): accessibility in support of materials research and design. Acta Crystallogr A 58:364–369
63. Grazulis S et al (2009) Crystallography open database an open-access collection of crystal structures. J Appl Crystallogr 42:726–729
64. Downs RT, Hall-Wallace M (2003) The american mineralogist crystal structure database. Am Mineral 88:247–250
65. Eyring H (1935) The Activated complex in chemical reactions. J Chem Phys 3:107–115
66. Laidler K, King C (1983) Development of transition-state theory. J Phys Chem 87 (15):2657–2664
67. Henkelman G, Jóhannesson G, Jónsson H (2000) Methods for finding saddle points and minimum energy paths. In: Schwartz SD (ed) Progress on theoretical chemistry and physics. Kluwer Academic Publishers, pp 269–300
68. Henkelman G, Uberuaga BP, Jónsson H (2000) A climbing image nudged elastic band method for finding saddle points and minimum energy paths. J Chem Phys 113(22):9901–9904
69. Yan X, Gouissem A, Sharma P (2015) Atomistic insights into Li-ion diffusion in amorphous silicon. Mech Mater. doi:10.1016/j.mechmat.2015.04.001
70. Malik R, Burch D, Bazant M, Ceder G (2010) Particle size dependence of the ionic diffusivity. Nano Lett 10(10):4123–4127
71. Arora P, White RE, Doyle M (1998) Capacity Fade Mechanisms and Side Reactions in Lithium-Ion Batteries. J Electrochem Soc 145(10):3647–3667
72. Vetter J, Novák P, Wagner MR, Veit C, Möller K-C, Besenhard JO, Winter M, Wohlfahrt-Mehrens M, Vogler C, Hammouche A (2005) Ageing mechanisms in lithium-ion batteries. J Power Sources 147(1–2):269–281
73. Christensen J, Newman J (2005) Cyclable lithium and capacity loss in Li-Ion cells. J Electrochem Soc 152(4):A818–830
74. Schmickler W, Santos E (eds) (2010) Interfacial electrochemistry, 2nd edn. Springer, Berlin
75. Nørskov JK, Rossmeisl J, Logadottir A, Lindqvist L, Kitchin JR, Bligaard T, Jonsson H (2004) Origin of the overpotential for oxygen reduction at a Fuel-cell cathode. J Phys Chem B 108(46):17886–17892
76. Lozovoi AY, Alavi A, Kohanoff J, Lynden-Bell RM (2001) Ab initio simulation of charged slabs at constant chemical potential. J Chem Phys 115(4):1661–1669
77. Filhol J-S, Neurock M (2006) Angew Chem Int Ed Elucidation of the Electrochemical Activation of Water over Pd by First Principles 45(3):402–406
78. Filhol J-S, Bocquet M-L (2007) Charge control of the water monolayer/Pd interface Chem. Phys Lett 438(4–6):203–207
79. Taylor CD, Wasileski SA, Filhol J-S, Neurock M (2006) First principles reaction modeling of the electrochemical interface: Consideration and calculation of a tunable surface potential from atomic and electronic structure. Phys Rev B 73(16):165402–1–16

80. Mamatkoulov M, Filhol J-S (2011) An ab initio study of electrochemical vs. electromechanical properties: the case of CO adsorbed on a Pt (111) surface. Phys Chem Chem Phys 13(17):7675–7684

81. Filhol J-S, Doublet M-L (2013) An ab initio study of surface electrochemical disproportionation: the case of a water monolayer adsorbed on a Pd (111) surface Catal Today 202:87–97

82. Lespes N, Filhol J-S (2015) Using the electrochemical dimension to build water/Ru (0001) phase diagram. Surf Sci 631:8–16

83. Dalverny A-L, Filhol J-S, Doublet M-L (2011) Interface Electrochemistry in Conversion Reactions for Li-Ion Batteries. J Mat Chem 21:10134–10142

84. Lespes N, Filhol J-S (2015) Using Implicit Solvent in Ab Initio Electrochemical Modeling: Investigating Li+/Li Electrochemistry at a Li/Solvent Interface. J Comp Theo Chem. 11 (7):3375–3382 doi:10.1021/acs.jctc.5b00170

85. Ando K, Hynes JT (1997) Molecular mechanism of Hcl acid ionization in water: Ab initio potential energy surfaces and Monte Carlo simulations. J Phys Chem B 101(49):10464–10478

86. Del Popolo MG, Lynden-Bell RM, Kohanoff J (2005) Ab initio molecular dynamics simulation of a room temperature ionic liquid. J Phys Chem B 109(12):5895–5902

87. Tomasi J, Mennucci B, Cammi R (2005) Quantum mechanical continuum solvation models. Chem Rev 105(8):2999–3093

88. Cossi M, Rega N, Scalmani G, Barone V (2003) Energies, structures, and electronic properties of molecules in solution with the C-Pcm solvation model. J Comput Chem 24(6):669–681

89. Jinnouchi R, Anderson AB (2008) Electronic structure calculations of liquid-solid interfaces: combination of density functional theory and modified Poisson-Boltzmann theory. Phys Rev B 77(24):245417-1-18

90. Andreussi O, Dabo I, Marzari N (2012) Revised self-consistent continuum solvation in electronic-structure calculations. J Chem Phys 136(6):064102-1-20

91. Steinmann S, Michel C, Schwiedernoch R, Filhol J-S, Sautet P (2015) Modeling the HCOOH/CO2 Electrocatalytic Reaction, When Details Are Key. Chem Phys Chem. 16 (11):2307–2311. doi:10.1002/cphc.201500187

92. Filhol J-S, Doublet M-L (2014) Conceptual surface electrochemistry and new redox descriptors. J Phys Chem C 118(33):19023–19031

93. Khatib R, Dalverny A-L, saubanère M, Gaberscek M, Doublet ML (2013) Origin of the voltage hysteresis in the CoP conversion material for Li-Ion batteries. J Phys Chem C 117:837–849

94 Meggiolaro D, Gigli G, Paolone A, Reale P, Doublet M-L, Brutti S (2015) Origin of the voltage hysteresis of MgH 2 electrodes in Lithium. J Phys Chem C 119:17044–17052

Multi-scale Simulation Study of Pt-Alloys Degradation for Fuel Cells Applications

G. Ramos-Sánchez, Nhi Dang and Perla B. Balbuena

Abstract Low-temperature fuel cells are one of the most promising systems for the transformation of fuels in an efficient, silent, and environmentally friendly manner. The requirements for the electrocatalyst are essentially three: the highest possible catalytic activity and the longest life cycle at the lowest cost. Sometimes, we can obtain one at expenses of the other. In this chapter, we review the simulation methods used in our group to study the degradation of catalysts for fuel cell applications: Density functional theory (DFT), classical molecular dynamics (CMD), Ab initio molecular dynamics (AIMD), and kinetic Monte Carlo (KMC). In the first part, we employ DFT, AIMD, and CMD to address the importance of the oxygen concentration on the surface of the catalysts and its influence on the "buckling" of Pt atoms and the role of the subsurface atoms. Then we analyze the temporal evolution of shape and composition of Pt/Ni nanoalloys by KMC simulations at various overall compositions and applied voltages. Finally, using DFT we study the effect that the presence of oxygen in the subsurface has on the buckling of Pt skin/PtCo structures by varying the oxygen coverage factor. The different methods and time scales used for the simulations permit us to fathom the factors governing the stability of electrocatalysts for fuel cells applications.

1 Introduction

In the last centuries, fossil fuels have helped mankind to obtain increasing levels of comfort and wellness. Unfortunately, the use of fossil fuels has also caused detrimental effects on the environmental global conditions. Now it is widely accepted that the global warming and climate change are caused directly by human activity, specially owed to CO_2 emissions generated by energy and transportation demands.

G. Ramos-Sánchez · N. Dang · P.B. Balbuena (✉)
Department of Chemical Engineering, Texas A&M University, College Station,
TX 77843, USA
e-mail: balbuena@tamu.edu

© Springer-Verlag London 2016   37
A.A. Franco et al. (eds.), *Physical Multiscale Modeling and Numerical
Simulation of Electrochemical Devices for Energy Conversion and Storage*,
Green Energy and Technology, DOI 10.1007/978-1-4471-5677-2_2

Despite many actions proposed by international organisms in the reduction of greenhouse gases, and promotion of conscious usage of the automobile, the CO_2 and CO emissions are continuously increasing to values that may lead to unthinkable consequences [1]. Renewable energies are promising alternatives to the use of fossil fuels: solar, wind, and sea are practically infinite and free energy sources if we find a cheap way to convert, storage, and transmit them. Solar energy, wind, and hydropower cannot be widely used without converting into electricity or chemical energy carriers. The favored chemical energy carriers are hydrogen and methanol because of their "easy" generation, respectively, from water and carbon dioxide, both lead to a drastic CO_2 reduction compared to the fossil-based concepts [2]. Low-temperature fuel cells (Polymer electrolyte fuel cells, PEFC) represent an environmentally friendly technology and are attracting considerable interest as a means of producing electricity by direct electrochemical conversion of hydrogen/methanol and oxygen into water and carbon monoxide.

There are, however, severe shortcomings on the present technologies, which need to be overcome to make low-temperature fuel cells economically attractive. One of the most important problems is related to the low rate of the cathodic reaction, the oxygen reduction reaction (ORR). Platinum has been widely used for this reaction, but due to kinetic limitations the cathodic overpotential losses amount to 0.3–0.4 V under typical PEFC operating conditions. Improvements have been made, especially by alloying or modifying the composition of the cathodic nanoparticle. The improvement in the ORR electrocatalysts of Pt-alloys has been ascribed to different structural changes caused by lattice mismatch producing decreased Pt–Pt distances and electronic or ligand effects in which a different environment causes the modification of the density of states responsible of the coupling with the oxygen molecule. Pt/Co and Pt/Ni catalysts in different structural configurations and proportions of its components have been proposed to be among the most active materials able to overcome the large energy barrier for the electrochemical reduction of molecular oxygen [3–7].

Tremendous improvements in power density and cost reduction of polymer electrolyte fuel cells have been achieved over the last two decades, which have brought PEFC systems close to the benchmarks that are critical for commercialization [8]; however, several other problems have arose, and these issues are mostly related to the time evolution of the electrode materials. Electrode durability is now one of the main shortcomings limiting the large-scale development and commercialization of this zero-emission power technology. The change in the microstructure is related to degradation/corrosion of the catalytic layers components: carbon and platinum-based nanoparticles, but also corrosion of other components is possible: membrane degradation and corrosion of the bipolar plates. It is widely reported that the electrochemical and microstructural properties of Pt and Pt-alloy electrodes evolve during the membrane electrode assembly (MEA) operation and in typical three electrode probes [9–11]. Degradation of the membrane has been related to radical attack coming from hydrogen peroxide produced in the two-electron oxygen reduction [12]; this is detrimental because it ultimately results in membrane pin-hole formation, which is the main cause of cell failure. The resultant structure after catalysts and substrate degradation eventually leads to the formation of cracks and in consequence to the separation of membrane and catalytic layer with final consequences such as increase in total resistance, decrease of cell voltage, increase of mass transport issues, decrease of the initially optimized 3D structures, etc.; all these phenomena drive unacceptable performance losses over time and eventually cell failure [13]. In this work, we focus on the degradation problems originated in the structure and composition of the catalytic material, particularly in the alloyed catalysts where the inclusion of a second or third alloying metal causes different durability properties than those of pure Platinum.

For nanoalloys in spite of the great deal of efforts set on the synthesis, it is very likely that the synthesized compounds appear very promising regarding size distributions, and other desirable features, once in the electrochemical media of fuel cells their structure changes rapidly; so different experimental and theoretical tools are used to evaluate the temporal evolution of the catalysts. Several mechanisms have been proposed to determine the instability of Pt nanoparticles (and alloys) in low-temperature fuel cells [14, 15]. The various proposed mechanisms are depicted in Fig. 1: (a) and (b) include the formation of ionic species that are then reduced and

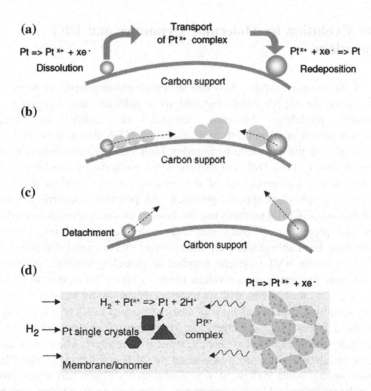

Fig. 1 Main mechanisms for Pt degradation in PEMFC environment. Reprinted with permission from Ref. [15]. **a** Growth via modified ostwald ripening. **b** Coalescence via crystal migration. **c** Detachment from carbon support. **d** Dissolution and precipitation in the ion conductor.

deposited in a larger nanoparticle, or they are reduced by the anodic hydrogen crossover in the membrane. The (b) and (c) mechanisms depend on the mobility of the nanoparticle which is directly related to the substrate–nanoparticle interaction, and to the substrate corrosion. The dissolution of metallic species is modified by the presence of an alloyed element that is more likely to be oxidized than platinum, but also is possible that the alloyed element provokes that the Pt in the surface changes its thermodynamic and kinetic tendency to be detached from the surface. In our research group, we have taken several routes to analyze the stability of metallic nanoparticles in PEFC environments. (1) Using the DFT approach, we investigate the substrate–nanoparticle interaction and its connection with reactivity [16, 17]; (2) Using kinetic Monte Carlo (KMC) techniques, we describe the main processes that are likely to occur in the electrochemical environment such as dissolution and diffusion of the alloy components, reaching simulation times in the order of days; in comparison with simulations of nanoseconds (CMD) or a few picoseconds (AIMD); (3) In a possible structure obtained by the KMC method, we use DFT to analyze the effect that the oxygen atoms in the subsurface have on the adsorbed atoms on the surface. In the two following sections, these approaches are addressed in more detail.

2 Time Evolution by Molecular Dynamics and DFT Simulations

Because of the strong coupling between different physicochemical phenomena, interpretation of the experimental observations is difficult, and analysis through mathematical modeling becomes crucial in order to establish microstructure-performance relationships, elucidate MEA degradation and failure mechanism, and, in the end, help to improve both PEFC electrochemical performance and durability [18]. Different mathematical methods are available to study specific time scales and properties of the system, i.e., the mathematical model includes the description of specific particles and processes occurring. Classical molecular dynamics (CMD) methods use the laws of classical physics to predict the structures and properties of molecules and molecular assemblies. Using this method, we have been able to study the oxidation of Pt(111) and PtCo alloys, using a canonical ensemble NVT (constant number of particles, volume, and temperature), at different oxygen coverage values using Lennard–Jones potentials for the Pt–Pt and Co–Co interactions with parameters fitted to Sutton–Chen potential energy. By assigning negative charges to adsorbed oxygen and positive charges to the Pt atoms in the two topmost layers, we simulated the environment found in a PEFC catalytic system at different electrochemical conditions. The atomic charges used for the simulation were directly obtained by DFT simulations (Bader Charges after optimization). After 900 ps, it was found that oxygen and water produce changes in the structure and local composition of the near-surface layers which may affect the activity and stability of the catalyst. Such changes are strongly dependent

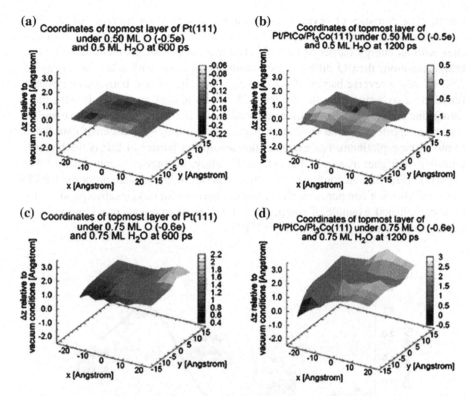

Fig. 2 Molecular dynamics snapshots of the Pt (**a** and **c**) and Pt/PtCo/PtCo₃ (**b** and **d**) under different oxygen coverage conditions and electric charge in oxygen. Adapted from Ref. [19]. Reproduced by permission of the PCCP Owner Societies

on the amount of adsorbed oxygen which enhances the segregation of cobalt to the topmost layer. In Fig. 2 (modified from reference [19]), it is possible to observe differences in the topmost layer in a fcc Pt(111) and Pt/PtCo/Pt₃Co at the same conditions of charge and oxygen coverage. It is observed that the separation of the topmost layer from the surface is higher in the alloyed PtCo system than in Pt. In the pure Pt system, the simulation leads to an almost planar structure (Fig. 2a) in comparison to the Pt/PtCo/Pt₃Co system where buckling of atoms is found and is enhanced by augmenting the oxygen coverage. These simulations also were able to confirm the experimental results of oxygen in the subsurface layers [20, 21] and the presence of more bonds Co–O than Pt–O; at oxygen coverage of 0.5 and 0.6 ML, most of the oxygen atoms remain adsorbed on the surface and Co–O bond is only twice more frequent than Pt–O bond. However, for 0.85 of oxygen coverage most of the oxygen atoms are absorbed deeply into the subsurface and the Co–O bonds are three times more frequent than Pt–O bonds.

The absorption of oxygen into the subsurface in pure metals and alloys has been identified as one of the first stages of the catalyst corrosion [22]. In a series of studies using electronic structure methods, analysis of the thermodynamic and kinetic characteristics of oxygen adsorption and absorption were reported [23–25]. A mechanism for oxygen absorption was found where O migrates from fcc to hcp sites with an energy barrier of 0.63 eV. The transition state geometry is with O in bridge position; then O diffuses into subsurface fcc site with a barrier energy of 2.22 eV and a reverse barrier of 0.7 eV (Fig. 3a). The reverse barrier changes as a function of coverage from 0.7 eV for 0.25 ML to almost zero for 0.11 ML, indicating that subsurface absorption is allowed by higher surface oxygen coverage. It was also reported that the absorption barriers in Pt-alloys are different than those found in pure platinum. For example, the absorption barrier in PtIr is the highest, almost 1 eV higher than in Pt, but in the PtCo alloy, the barrier is only a half of that in pure Pt (Fig. 3b). According to this study, a series of alloys were proposed with Ir as a third alloying component with the highest barriers for oxygen absorption, being an accomplished example of computational design of materials providing guidelines in order to enhance the stability [25].

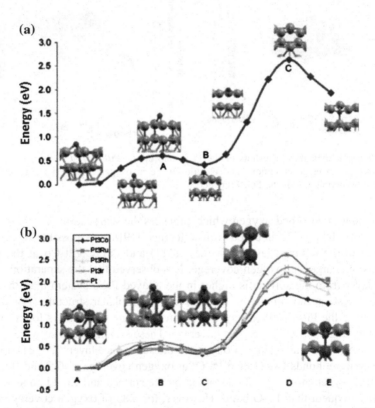

Fig. 3 Energetic barriers for the absorption of oxygen in **a** pure Pt fcc and **b** several PtM alloys where M is a transition metal, adapted from [23–25] (Pt barriers are the same in **a** and **b**). Reprinted with permission from J. Phys. Chem. C., 2007, 111, 9877-9883. Copyright 2007 American Chemical Society; J. Phys. Chem. C, 2007, 111, 17,388-17,396. Copyright 2007 American Chemical Society; J. Phys. Chem. C, 2008, 112, 5057-5065. Copyright 2008 American Chemical Society

3   Time Evolution of PtM Alloys by KMC Methods

Monte Carlo refers to a broad class of algorithms that solve numerical problems through the use of random numbers [26]. A specific coarse-grained Monte Carlo method is that intended for systems evolving dynamically from state to state referred as KMC and widely used for surface reactions, electrochemical systems, and growth processes [27–31]. The dynamical evolution of a system can be studied by molecular dynamics in which one propagates the classical equations of motion forward in time. To do so, one requires suitable interatomic and intermolecular potentials and a set of boundary conditions. The behavior of the system emerges naturally, requiring no intuition or additional input from the user; therefore, the accuracy of the result depends on the parameterization of the atomic forces. A serious limitation is that accurate integration requires time steps short enough to resolve the atomic vibrations. Consequently, the total simulation time in general is limited to much less than one microsecond, usually in the order of tens of nanoseconds. However, the KMC method exploits the fact that the system evolves through diffusive jumps from one state to another, rather than following the vibrational trajectory; consequently, KMC can reach vastly longer scales [32]. Assuming that we know what the possible microscopic events are, we can use transition state theory (TST) to compute the rate constant for each event. Thus, using high-quality TST rates for all possible events, KMC simulations can, in principle, be made as representative of the real process as the results from MD simulations.

Recently, we have reported a KMC algorithm that is able to reproduce the experimental dealloying of Pt-based nanoparticles typically used as oxygen reduction reaction catalyst; the details of the simulation can be found in the original publication [33]. Here, we present a brief introduction to the parameters and reactions used in the original paper. A 3D nanoparticle with metallic structure and lattice parameters optimized using DFT simulations for pure metals and alloys surrounded by a layer of electrolyte is used as starting point. Then each species occupying a site may participate in two reactions: diffusion and dissolution, both described by the Arrhenius equation. The activation energy for the Arrhenius equation includes a bonding energy that depends on (1) the coordination of the atoms and the bonding energy, which was calculated by DFT simulations, and (2) the electric potential of the system by a modification of the Butler–Volmer equation [33] taking into account the transfer factor (having a value of 0.5 for dissolution and 0 for diffusion) and the standard potential for the dissolution of the different metallic atoms. The bonding parameters between Pt and electrolyte were tuned through KMC simulations using the surface diffusion coefficient obtained from experiments and are kept independent of potential.

Using the KMC algorithm, it was possible to study the synthesis of hollow nanoparticles resulting from removal of the alloy core through dealloying using the Kinkerdall effect [34], and the synthesis of porous nanoparticles by selective dealloying of the less noble component. The importance of hollow and porous nanoparticles resides on their specific area; the specific mass activity is twice higher for the hollow nanoparticles in comparison to the single core–shell structures. For the formation of hollow nanoparticles in a range of particle diameters between 5 and 8 nm, it was found that an applied potential of 12 V is necessary to enhance the diffusion of core species. This is because the core atoms are strongly bonded to their neighbors and these atoms must follow a high-energy path by exchanging positions directly with other metal atoms. In order to have a lower diffusion barrier, the presence of vacancies creates a low-resistance path facilitating the migration of the non-noble metal atoms to the surface where they dissolve. Figure 4 illustrates the required applied voltage for the formation of porous and hollow nanoparticles, which has been discussed previously for nanoparticles with diameter smaller than 15 nm [35]. It is possible to observe ranges of potential in which the structure of the nanoparticle changes drastically, lower potentials lead to core–shell structures, intermediate values to porous, and very high potentials to hollow nanoparticles.

The size of the nanoparticles normally used in fuel cells is ∼5 nm. For this reason, we next are interested in the evolution of nanoparticles of this size and potential ranges common to fuel cell conditions 0–1.2 V and with the presence of oxygen. The nature and concentration of species present on the surface of the nanoparticles are functions of the applied potential. Many species can be interacting such as OH, O, O_2, H_2O, and H_2O_2 in different concentrations and their influence can greatly affect the evolution of the nanoparticle. However, inclusion of all these factors makes the simulation much more complex. Thus, at this stage we focus on the influence of oxygen at two potentials 0.9 V with 20 % of oxygen on the surface and 1.2 V with 80 % of oxygen on the surface and 20 % of oxygen in the subsurface [36, 37].

A 5-nm-diameter nanoparticle $Ni_{0.75}Pt_{0.25}$ initially with a random configuration was analyzed by the KMC algorithm. The initial configuration and those after 10^5 and 10^6 s are depicted in Fig. 5. The applied potential is 0.9 V NHE (Normal hydrogen electrode) and oxygen is allowed to interact with the surface forming a

Fig. 4 Structures formed in the Pt–Ni alloy according to the applied potential. Adapted from Ref. [33]

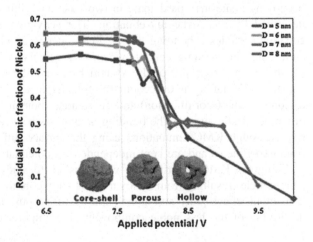

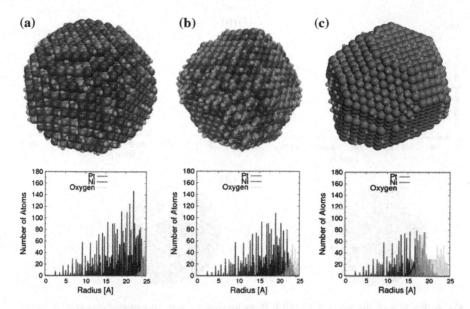

Fig. 5 Evolution of the random alloy as function of time at 0.9 V and 20 % of oxygen coverage. *Top* snapshots (*Blue*, *gray*, and *red spheres* are Ni, Pt, and oxygen, respectively). *Bottom* radial atomic distribution. **a** Initial random configuration, **b** 10^5 s, and **c** 10^6 s (Color figure online)

constant coverage on fcc sites. We emphasize that this alloy has a high Ni percentage at a low potential, in comparison to the potential used for the formation of hollow nanoparticles. The initial random configuration snapshot (Fig. 5a) shows a high concentration of Ni atoms on the surface. After 10^5 s, many Ni atoms dissolve and Pt atoms tend to occupy surface positions, after 10^6 s all the atoms in the surface are Pt and after having reached this configuration, only slight changes are observed. As a consequence of the dissolution of Ni atoms, the nanoparticle gets smaller, the initial radius was 2.5 nm, and at the end of the simulation the nanoparticle has a 2.2-nm radius. These results are in agreement with the previously reported conclusions that at potentials below 7 V the nanoparticle acquires a core–shell configuration. However, in this case the potential is very low and after ~ 24 h the final configuration shows a total rearrangement of the structure with all the Pt atoms covering the nanoparticle.

The main reason to have a Pt shell with a Pt/Ni core is the dissolution of Ni atoms, rather than diffusion of Pt atoms to the surface. The $Ni_{0.75}Pt_{0.25}$ nanoparticle experiences dissolution of 10 % of the Ni atoms at the very beginning of the simulation and continues all over until the Pt shell is formed. On the other hand, the fraction of Pt atoms that are being dissolved seems to be almost constant (Fig. 6). In order to analyze the influence of different proportions of the alloy constituents, we simulated the $Ni_{0.05}Pt_{0.95}$ and $Ni_{0.25}Pt_{0.75}$ alloys at the same conditions. Figure 6 illustrates that the alloy with 25 % of Ni has almost the same behavior than that with 5 % Ni at short time periods but it reaches faster a constant Ni composition.

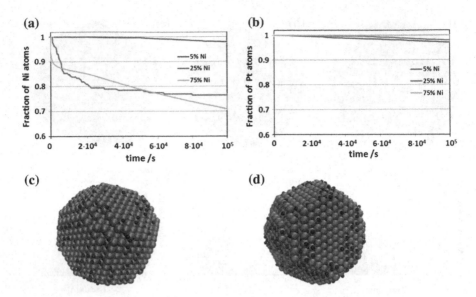

Fig. 6 Fraction of the atoms **a** Ni and **b** Pt as function of time in nanoparticles with different compositions. Snapshots of the final configuration (10^5 s) of the alloys with **c** 25 % Ni and **d** 5 % Ni

The alloy with 5 % loses only a few percent of surface Ni atoms, and then it acquires a more stable configuration that contains a full monolayer of Pt atoms on the surface. The same occurs with the alloy with 25 % Ni. The final configurations for the alloys with 25 and 5 % of Ni are depicted in Fig. 6c and d, respectively; these configurations are attained at 10^5 s in comparison with the alloy with 75 % Ni that requires 10^6 s to achieve the same configuration.

Next, we are interested in the evolution of the same nanoparticle as a function of time at different potentials. The applied potential affects the systems in many ways; it could provide a driving force for the electrochemical dissolution of one of the alloy components or both. It also affects the velocity at which the surface reactions occur (oxygen reduction on the cathode and hydrogen oxidation on the anode of a fuel cell), so it is directly related to the concentration of the surface species. It has been found by both theoretical and experimental techniques that the concentrations of oxygen on the surface and in the subsurface are functions of the potential; therefore, simulations of the nanoparticle at different potentials should include the correct representation of the oxygen concentration. We chose two potentials: (1) 0.9 V at which there is no subsurface oxygen and the concentration of oxygen is 22 % on the hollow fcc sites and (2) 1.2 V at which the concentration of oxygen on the surface is 90 % and the subsurface concentration is 34 %. The adsorption only was allowed on hollow fcc sites on the surface and absorption was allowed in octahedral sites of the subsurface. Figure 7 depicts the time evolution of a nanoparticle with the same diameter (5 nm) at different initial Pt:Ni ratios and at different electrochemical potentials.

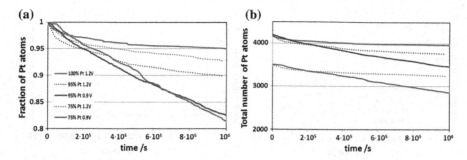

Fig. 7 Time evolution of the platinum atoms in a 5-nm alloy Pt–Ni with the percentage of Pt and potential applied indicated in the legend. **a** Fraction of Pt atoms (with respect to the initial number of Pt atoms) in the nanoparticle, and **b** total number of Pt atoms in the nanoparticle

The nanoparticle composed of 100 % of Pt at 1.2 V presents the lowest dissolution of Pt atoms. At the very beginning of the simulation ($0–1 \times 10^5$ s), the nanoparticles with 75 and 95 % of platinum at 0.9 V show lesser dissolution of Pt atoms, which is related to the preferential dissolution of the Ni atoms. After this time, the pure platinum nanoparticle attains a stable configuration and the number and fraction of Pt atoms is kept almost constant, even at times as higher as 1×10^8 s; the nanoparticle fraction of atoms does not change less than 8 %. The total number of atoms and the fraction of atoms as functions of potential follow two different trends. At times lower than 2×10^5 s, the nanoparticles at 0.9 V independently of their composition lose a smaller amount of Pt atoms, which is directly related to a higher driving force for dissolution of Ni atoms and consequently higher dissolution. However, after 2×10^5 s the system follows the opposite trend, with nanoparticles under a lower potential showing a higher dissolution of Pt atoms. This effect occurs despite of the lower applied potential and that at 1×10^5 s at 0.9 V (Fig. 6c, d) the nanoparticle surfaces are composed practically only by Pt atoms. Therefore, at 0.9 V the dissolution of Pt is enhanced by the presence of Ni in the second layer. However, at 1.2 V the higher oxygen concentration allows the formation of an oxide layer that inhibits the dissolution of Pt atoms, but the alloyed nanoparticle presents higher dissolution of Pt atoms than the nanoparticle composed of 100 % of platinum. In general, the dissolution of Pt atoms is enhanced by the presence of an alloying component (Fig. 8).

The fraction of dissolved Ni atoms is higher in comparison to the Pt atoms; at the same elapsed time, the dissolution of Ni atoms reaches values as high as 50 % of the total number of atoms. At the very beginning of the simulation, i.e., when the nanoparticle is allowed to interact with the electrochemical media, more than the 20 % of the Ni is dissolved at 0.9 V; when the nanoparticle is at 1.2 V, only the 10 % of the atoms are lost. The same scenario is observed at higher potentials: the oxygen atoms form an oxide layer that prevents the dissolution of the Ni atoms. As can be observed in Fig. 9, the high O concentration is responsible for the formation of an oxide layer that avoids the detachment of both Pt and Ni atoms.

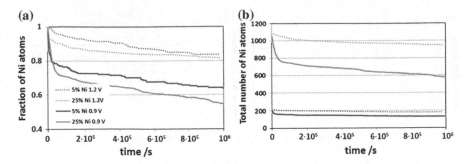

Fig. 8 Time evolution of Ni atoms on the PtNi nanoparticles with the composition and at the potentials indicated in the legend. **a** Fraction of Ni atoms (with respect to the initial number of Ni atoms) as a function of time and **b** total number of Ni atoms as a function of time

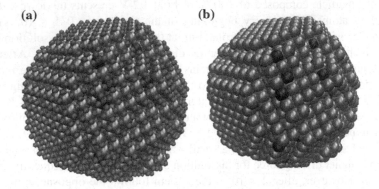

Fig. 9 Snapshot of a $Pt_{0.95}Ni_{0.05}$ at 1.2 V after 1×10^6 s of KMC simulation. *Gray, blue, red*, and *purple* spheres represent platinum, nickel, oxygen on the surface, and oxygen in the subsurface, respectively. **a** Structure showing oxygen on the surface, **b** surface oxygen is not shown to better visualize the surface and subsurface composition (Color figure online)

Note that more Ni atoms are observed on the surface in contrast with Fig. 6, where most of the atoms on the surface are platinum.

The results here obtained have important implications: (1) The formation of a metallic alloy in order to enhance the ORR kinetics causes enhanced degradation owed to the dissolution of the less noble alloyed metal but also to the enhanced dissolution of Pt in comparison to the pure nanoparticle. (2) The formation of an oxide layer at 1.2 V avoids the degradation of the nanoparticle; however, the usual range of operation of this nanoparticles is 0.6–1.0 V which imply that dissolution of the components in this range is expected. (3) The subsurface concentration of oxygen has been kept constant, and it is essential for improved results to investigate the mobility and effect that these atoms may have on the detachment of Pt and alloyed metals. And (5) Of special importance is the analysis of dynamic potential, i.e., what would be the effect of potential steps or potential sweep on the stability of the nanoparticles. This will be reported elsewhere.

4   Degradation of PtCo Skin

In the previous section, the position and charge of oxygen in the subsurface was kept constant; however, those changes should be analyzed along with the effect on the structure of the surface. Thus, in this section the evolution of a Pt/PtCo skin system is analyzed using DFT after the absorption of oxygen in the subsurface. The system used for the simulations consists of a 2 × 2 fcc slab containing five layers separated by a 12 Å vacuum. The three first layers were allowed to relax, while the last two are kept fixed. The first layer has only Pt atoms, the second only Co atoms, and from the third to fifth a 1:1 composition. This system is the one previously reported by our group as the most stable obtained by DFT simulations, as well as confirmed by the KMC simulated experiments [38]. We have analyzed the influence of the O coverage factor on the stability of the PtCo system. We probed first the adsorption energy of oxygen either in the surface or in the subsurface for the formation of a 0.25 ML. In previous studies, it was found that the preferential sites for oxygen adsorption are the three-coordinated ones over atop and bridge sites [39–41]. In this work, we use the fcc and hcp sites for oxygen adsorption on the surface and the sub-fcc, sub-hcp, and sub-tet, all of them are depicted in Fig. 10. In the hcp site, the oxygen is above a Co atom in the second layer; in the fcc site, on the other hand, the oxygen is in line with a Pt atom in the third layer. The hcp-sub and fcc-sub sites are the counterparts of the hcp and fcc sites on the surface, the hcp-sub site is in a tetrahedral hollow, and the fcc-sub site is in an octahedral hollow. The tet-sub site is also in a tetrahedral hollow but in this case the interaction is with three cobalt atoms in the second layer and one platinum atom in the first

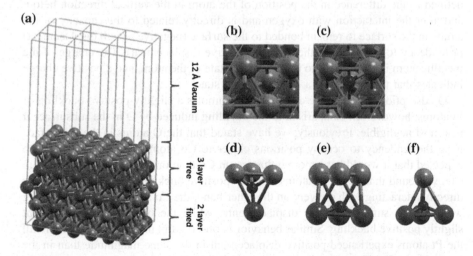

Fig. 10 a Slab model. Sites for oxygen adsorption: **b** hcp, **c** fcc; and for oxygen absorption in the subsurface: **d** hcp-sub, **e** tet-sub, and **f** fcc-sub. *Blue, gray,* and *red* spheres represent Co, Pt, and O atoms, respectively (Color figure online)

Table 1 Energetic and some geometrical values of O adsorption on the surface and absorption in the subsurface sites

		hcp	fcc	hcp-sub	tet-sub	fcc-sub
E_{ads}/eV	RPBE	−2.38	−2.57	+0.12	−1.99	−1.156
	PBE [43]	−2.94	−3.15			
	Pd@Co (PW91) [44]	−2.5				
Distance/Å	Pt–O	2.12	2.10	1.98	2.11	2.68
	Pt–O [43]	2.1	2.1			
	Co–O			1.80	1.86	1.82

layer, while in the hcp the interaction is with one cobalt atom in the second layer and three Pt atoms in the first layer. The adsorption energy is reported in Table 1, compared with results from other publications.

The adsorption energy calculated for the formation of 0.25 ML with the RPBE [42] exchange correlation functional is weaker in comparison with the PBE functional; however, the same trends are obtained with both exchange correlation functionals. The O adsorption energy in the surface is higher than the absorption in the subsurface, and interestingly the absorption in the hcp-sub site presents positive adsorption energy. The most stable site for adsorption on the surface is the fcc with very similar values for the hcp site, and the most stable site for the absorption in the subsurface is the tet-sub site. In any of the subsurface sites in the subsurface, the O atom has a tendency to stay closer to the Co than to the Pt atoms.

It is interesting to analyze what are the physical consequences of the oxygen adsorption on each interaction site. We focus on the vertical displacement of the surface atoms (buckling) and the charge of the atoms. The vertical displacement is defined as the difference in the position of the atom in the vertical direction before and after the interaction with oxygen and is directly related to the capacity of the atoms in the surface to remain bonded to the surface (negative or null displacement) or tendency to escape from the surface (positive displacement). The charge of the metallic atom is also related to the oxidation state of the atom, and a positive charge indicates that the atom is in a more oxidized state.

O adsorption, either in fcc or hcp sites, promotes a slightly positive buckling of Pt atoms; however, in comparison to the buckling induced by O in the subsurface it is almost negligible. Previously, we have stated that the O atoms in the subsurface have the tendency to occupy positions closer to Co atoms, and in this sense we expected that it would have more influence on Co than on Pt atoms. In the hcp-sub site, we found that all the Pt atoms suffered positive buckling, especially the atoms directly interacting with oxygen; on the other hand, the Co atom directly interacting with oxygen suffers negative displacements, while the other Co atoms present slightly positive buckling. Similar behavior is observed in the fcc-sub site where all the Pt atoms experienced positive displacement in the same magnitude than in the hcp-sub site. The Co atoms directly interacting with O do not change their positions but the one that is not directly interacting suffers a positive displacement of the

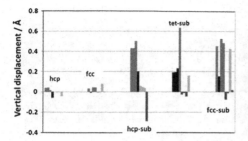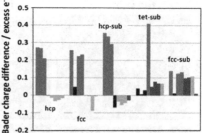

Fig. 11 Vertical displacement and Bader charge of surface and subsurface atoms. *Red bars* represent Pt atoms directly interacting with oxygen; *black bars* represent Pt atoms not in direct contact with oxygen. *Blue bars* represent Co atoms directly interacting with oxygen; *green bars* represent Co atoms not directly in contact with oxygen (Color figure online)

same magnitude than those of the Pt atoms. Interacting in the tet-sub site, the Pt atom directly above the oxygen in the subsurface presents the highest vertical positive displacement, but also the other Pt atoms not directly interacting with oxygen, whereas the Co atom that is not in contact with oxygen suffers a positive vertical displacement (Fig. 11).

In the surface of the clean PtCo skin, the Co atoms have positive charge (about +0.3 e/atom), while the Pt atoms in the deepest layers have negative values (−0.4 e/atom); however, the Pt atoms at the surface have a lesser negative value around −0.22 e/atom. After O adsorption if the change on the charge of platinum atoms is greater than 0.22, the atom changes to a slightly oxidized state. Most of the different O positions either in the surface or subsurface lead to at least one Pt atom in an oxidized state. Although expected, it is worth to mention that when O is on the surface the Pt atoms directly interact with it, and become oxidized, while the no interacting atom remains with partial negative charge. In the hcp-sub site, all the Pt atoms have a more drastic change in charge becoming more oxidized, while the Co and Pt atoms not in contact with oxygen gain electronic charge. In the tet-sub site, only the Pt atom directly interacting with oxygen lose electronic charge; however, all the other Pt or Co atoms also lose charge but do not acquire partial positive charge. Finally, in the fcc-sub site Pt and Co atoms directly interacting with oxygen lose electron density; this site is where the Co atoms lose the highest charge density, which already was positive. Summarizing, the interaction with oxygen leads to a positive charge in the Pt atoms, and the Co atoms are also affected but in a lesser degree even if the oxygen is in the subsurface and closer to the Co than to the Pt atoms; however, the charge of Co atoms was already even more positive than those of any Pt atom. Energetically, the least favorable site on the subsurface is the hcp-sub site in which the Pt atoms are highly oxidized and the Co atoms gain charge, and the most stable is the tet-sub site with one Pt atom bearing high positive charge and a large positive vertical displacement.

After considering the consequences of the O adsorption on the surface and absorption in the subsurface, we want to analyze what is the influence of O in the

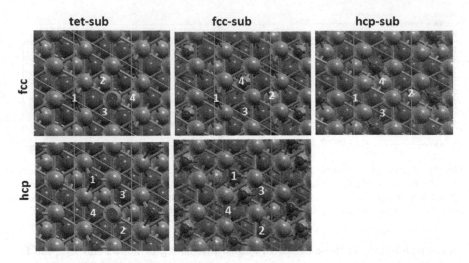

Fig. 12 Sites for adsorption of the four atoms on the surface, the column labels indicate the O site in the subsurface, and the row labels the position for O on the surface. The numbers indicate the order, which were allowed to adsorb; *1* = 0.25 ML, *2* = 0.5 ML, *3* = 0.75 ML, *4* = 1 ML

subsurface in both the adsorption energy of oxygen on the surface from 0.25 to 1 ML and the changes in vertical displacement and charge of the atoms.

Previous calculations performed by our group have shown that the energetic barrier for the absorption of oxygen is lower in the case of PtCo alloys (1.39 eV) in comparison to pure platinum (2.22); but also the energetic barrier for the reverse process is lower in PtCo (0.23 eV) than in Pt (0.7 eV) [25]. In order to allow the absorbed oxygen to desorb, the positions of the oxygen atoms on the surface (Fig. 12) were kept the farthest away from the absorbed oxygen atom during the process of completing the full coverage. In none of our simulations, the absorbed oxygen was desorbed from the subsurface, but the oxygen atom on the position with the lowest energy (hcp-sub) moved to the highest one (tet-sub), as in the case of the absorption in the hcp-sub site with adsorption in the hcp site. In this site, for all coverage values the oxygen in the subsurface changed to the tet-sub site. In all other combinations, the absorbed oxygen does not move from its original position as shown in Fig. 13, which illustrates the structural evolution of the slab as a function of the coverage, for every location of O in the subsurface and O adsorbed on the surface on the fcc sites. For comparison, we show the structural evolution of the system without oxygen in the subsurface as the oxygen coverage on the surface is changed.

The evolution of the system as a function of the surface oxygen coverage without oxygen in the subsurface is relatively simple. There are no major changes, and only at 0.75 ML the structure shows buckling but not as extreme as when the oxygen has been absorbed. For the system with oxygen located in the tet-sub position of the subsurface, the Pt atom above O is observed to protrude slightly above the surface. The vertical displacement is enhanced as the coverage augments,

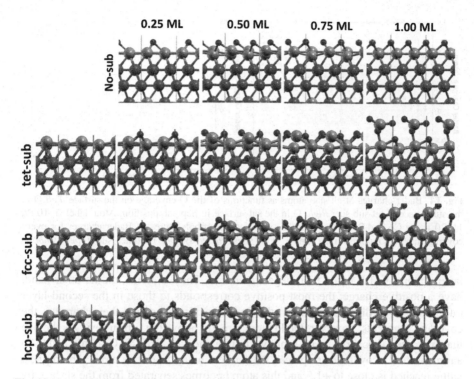

Fig. 13 Snapshots of the optimized structures at different oxygen coverage values with oxygen in the subsurface (2nd, 3rd, and 4th row) and without oxygen on the subsurface (1st row). *Gray, blue,* and *red spheres* represent platinum, cobalt, and oxygen atoms, respectively (Color figure online)

not for all the atoms but only for that atom originally located above the O; even if the Pt–O has been broken and the oxygen remains close to the Co atoms, the Pt atom continuously moves away from the surface. The final vertical displacement of the Pt atom with O on the surface at the fcc positions is 2.77 Å and with oxygen in hcp positions is 3.19 Å. All other atoms including Co and O remained in their same sites. Similar behavior is observed when O is in the fcc-sub site; at 1 ML coverage, one of the Pt atoms is bonded to the absorbed oxygen and another is detached from the surface, and the final vertical displacement of the Pt atom is 3.21 Å. The scenario changes when O is in the hcp-sub site. In that case, the O atom in the subsurface keeps linked to three Pt atoms. Surface detachment is observed even for very low O coverage factors but for all atoms in the same magnitude and no detachment is observed at full coverage; the three atoms moved from their original position but only 0.8 Å which is a minimum displacement in comparison to those induced by O in the fcc-sub and tet-sub sites.

In the charge analysis reported above, it was found that the Pt atom located directly above of the subsurface tet-sub oxygen has a net positive Bader charge, while the other Pt atoms in the first layer (atoms 1–4) retain their negative charge (water blue symbols in Fig. 14 left), and the Pt atoms in the third, fourth, and fifth

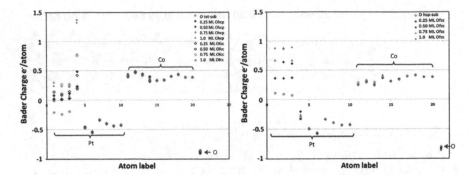

Fig. 14 Bader charges of all slab atoms as functions of the O coverage on the surface. *Left* O in the subsurface in tet-sub site; *Right* O in the subsurface in hcp-sub position. Atom label: 1–10 Pt; 11–20 is the Co and 21 is the O in the subsurface. Symbols represent Bader charges for every coverage having O atoms on the surface in fcc sites (*closed symbols*) and hcp sites (*open symbols*)

layers (atoms 6–10) retain their negative charge. The Co atoms, on the other hand, have a positive charge; the most positive corresponds to those in the second layer (atoms 11–14). All the Pt atoms in the first layer (1–4) increase their charge as the coverage augments; however, the Pt atom directly above the subsurface O is the one that increases its charge more rapidly, despite the fact that the O atoms added to augment the O coverage are placed the farthest away from this atom. The highest value reached is close to +1.5 and this atom becomes separated from the surface by more than 3 Å, clearly suggesting that this atom is ready to dissolve in the electrolyte or move to sites where it is able to receive electrons. The Co atoms in the second layer interestingly do not change their charge value neither their positions. During absorption of the O atom in the subsurface, the Co atoms do not change their charge value; therefore, in all the study the Co atoms are slightly affected by the subsurface atom and by the changes in the coverage.

The changes in the system charges with hcp-sub O are different. The three Pt atoms bonded to the subsurface O augment their charge value roughly in the same magnitude as the coverage increases; the maximum charge value reached in this site is lesser than +1 for all the atoms. The Co atoms in the second layer have lesser positive charges in comparison to the tet-sub site. In all the systems with subsurface O, the charges and positions of atoms in the second layer and deeper remain unaffected. In the system without subsurface O, the situation is totally different: the maximum charge reached by the Pt atoms in the fully oxygenated surface is +0.38 and that of the Co atoms is +0.3. We then can state that once O is in the subsurface despite the low barrier for desorption, the atom remains in the subsurface, and it eventually will lead to the detachment of the Pt atom directly above it. By keeping lower oxygen coverage, lesser than 0.5 ML, one ensures that the vertical displacement of the atom and its charge attains values similar to the system without subsurface O.

The adsorption energy was calculated as follows:

$$E_{ad} = E_{slab-O} - (E_{slab} - E_O) \qquad (1)$$

where E_{ad} is the adsorption energy, E_{slab} is the energy of the clean system, E_{slab-O} is the energy of the system with oxygen adsorbed either on the surface or in the subsurface, and E_O is the energy of the free oxygen atom. For subsequent adsorption steps, i.e., to augment the coverage, the reference energy (E_{slab}) is changed by the energy of the slab with a lesser oxygen atom and calculated in the same manner. For example, to calculate the E_{ad} for 0.75 ML, we use $E_{ad0.75ML} = E_{0.75ML} - (E_{0.50ML} - E_O)$. The relative energy is calculated as follows:

$$E_{rel} = E_{slab-nO} - (E_{slab} - nE_O) \qquad (2)$$

where E_{rel} is the energy relative to the clean surface, $E_{slab-nO}$ is the slab with n oxygen atoms either in the subsurface or on the surface, and nE_O is n times the energy of the free oxygen.

At zero coverage, the adsorption energy on the surface is stronger than the absorption in the subsurface; however, once O is in the subsurface the relative energy of the system with subsurface O is higher than that of the system without oxygen absorbed. Therefore, there is a thermodynamic driving force for one of the oxygen atoms on the surface to go to the subsurface; however, as previously mentioned the barrier for the atom to move to the subsurface is high at low coverage, and at high coverage the system stabilizes subsurface atomic oxygen from energetic and kinetic point of view [24]. The binding energy of the system with subsurface O is higher when the atom is in the tet-sub site at 0.25 ML, but at 0.5 and 0.75 ML the fcc-sub site has the highest binding energy, and at 1 ML they have the same binding energy; however, both the tet-sub and fcc-sub sites have similar values. The hcp site yields the weakest binding energy and in fact the site easily changes to tet-sub or fcc-sub site. In the full coverage situation, the hcp site without subsurface O yields stronger binding than the hcp-sub site indicating that this system has low probability to subsist changing either from the hcp-sub to fcc-sub site or desorbing the oxygen from the subsurface. The analysis of the total energy of systems with and without subsurface O could lead to misleading conclusions as the number of atoms is different in each system; however, if the relative energy of the system without subsurface O (red lines in Fig. 15) is moved 0.25 ML to the left, the same conclusions are reached: the system with subsurface O is more stable than the one without it.

The adsorption energy has been used as a descriptor of the catalytic activity, and most of the studies for the ORR on metallic surfaces have been tested in a 0.25 ML system [45], including that higher coverage is risky owed that the van der Waals interactions will play a more important role between adjacent molecules. Anyway the different values will give an idea of the changes in catalytic activity. The hcp-sub site is the system with the lowest binding energy. At all coverage situations, the binding energy in this site is the weakest one and possibly this situation

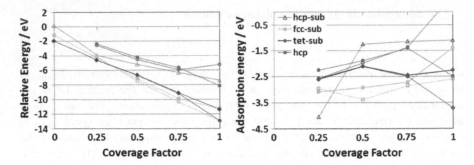

Fig. 15 Energetic values for oxygen adsorption as function of O coverage in fcc sites (*closed symbols*) and hcp sites (*open symbols*) with subsurface O in the positions indicated in the legend (legend is the same for both graphs) and without subsurface O (*red symbols*). Lines are drawn as guide for the eye. (Equations are explained in the main text). The legend indicates the oxygen in the subsurface. Two lines are drawn with the same color, one for the adsorption of oxygen in the fcc sites (*closed symbols*) and in the hcp sites (*open symbols*) (Color figure online)

never takes place. The binding energy is very similar in the tet-sub site to the situation without subsurface O. The binding energy in the systems without subsurface O tends to diminish as the coverage is increased, as expected due to higher interactions between adjacent molecules. The binding energy even reaches positive values at the full coverage system in the fcc sites. In the hcp site, the behavior is different because the surface O changes from hcp to top or bridge positions. In the systems with subsurface O, as one of the metal atoms moves away from the surface, the interactions between O atoms are weaker and in consequence the adsorption energy maintains the same values.

5 Concluding Remarks

A multi-scale modeling approach shows new insights regarding oxidation effects on alloy surfaces and nanoparticles. Combined DFT–classical MD simulations illustrate the time evolution of the catalyst surface as the oxygen coverage increases. The observed buckling is in agreement with previous DFT results where we investigated the adsorption and absorption of oxygen in alloy surfaces. Kinetic Monte Carlo simulations are used to determine the effect of an alloy on Pt dissolution under an electrochemical potential corresponding to the onset of surface oxidation. It is found that before the surface oxidation starts, first the dissolution of the non-noble metal takes place, and it is followed by the formation of a Pt shell on the surface of the nanoparticle. However, after the onset of surface oxidation, the oxidation film passivates the surface and stops metal dissolution; further research should be done on the effect of this passivating film on the catalytic activity. Finally, DFT calculations are utilized to understand the O-induced degradation of Pt–Co alloys at various coverage values. When there is no oxygen in the subsurface, structural changes are observed only at high surface coverage (>0.75 ML of O). In contrast, significant structural changes are detected when O is in the subsurface. Although buckling and surface reconstruction are observed for O absorbed in any of the subsurface sites, the most dramatic correspond to O absorbed in tet-sub and hcp-sub sites. The Co atoms in the second layer interestingly do not change their charge value or their positions. During absorption of the O atom in the subsurface, the Co atoms do not change their charge value; therefore, in all the study the Co atoms are slightly affected by the subsurface atom and by the changes in the coverage. It should be interesting to determine the mechanism for the Co atoms to move from the subsurface where they interact strongly with oxygen to the surface for their further dissolution.

References

1. Song CS (2006) Global challenges and strategies for control, conversion and utilization of CO_2 for sustainable development involving energy, catalysis, adsorption and chemical processing. Catal Today 115:2–32
2. Specht M, Staiss F, Bandi A, Weimer T (1998) Comparison of the renewable transportation fuels, liquid hydrogen and methanol, with gasoline-energetic and economic aspects. Int J Hydrogen Energy 23:387–396
3. Stamenkovic V, Schmidt TJ, Ross PN, Markovic NM (2002) Surface composition effects in electrocatalysis: Kinetics of oxygen reduction on well-defined Pt_3Ni and Pt_3Co alloy surfaces. J Phys Chem B 106:11970–11979
4. Xiong LF, Manthiram A (2004) Influence of atomic ordering on the electrocatalytic activity of Pt-Co alloys in alkaline electrolyte and proton exchange membrane fuel cells. J Mater Chem 14:1454–1460
5. Vinayan BP, Nagar R, Rajalakshmi N, Ramaprabhu S (2012) Novel platinum-cobalt alloy nanoparticles dispersed on nitrogen-doped graphene as a cathode electrocatalyst for PEMFC applications. Adv Funct Mater 22:3519–3526
6. Cui C, Gan L, Li H-H, Yu S-H, Heggen M, Strasser P (2012) Octahedral PtNi nanoparticle catalysts: exceptional oxygen reduction activity by tuning the alloy particle surface composition. Nano Lett 12:5885–5889
7. Loukrakpam R, Luo J, He T, Chen Y, Xu Z, Njoki PN, Wanjala BN, Fang B, Mott D, Yin J, Klar J, Powell B, Zhong C-J (2011) Nanoengineered PtCo and PtNi catalysts for oxygen reduction reaction: an assessment of the structural and electrocatalytic properties. J Phys Chem C 115:1682–1694
8. Gasteiger HA, Kocha SS, Sompalli B, Wagner FT (2005) Activity benchmarks and requirements for Pt, Pt-alloy, and non-Pt oxygen reduction catalysts for PEMFCs. Appl Catal B Environ 56:9–35
9. Xie J, Wood DL, Wayne DM, Zawodzinski TA, Atanassov P, Borup RL (2005) Durability of PEFCs at high humidity conditions. J Electrochem Soc 152:A104–A113
10. Ramos-Sanchez G, Solorza-Feria O (2010) Synthesis and characterization of Pd0.5NixSe (0.5-x) electrocatalysts for oxygen reduction reaction in acid media. Int J Hydrogen Energy 35:12105–12110
11. Borup RL, Davey JR, Garzon FH, Wood DL, Inbody MA (2006) PEM fuel cell electrocatalyst durability measurements. J Power Sources 163:76–81

12. Iojoiu C, Guilminot E, Maillard F, Chatenet M, Sanchez JY, Claude E, Rossinot E (2007) Membrane and active layer degradation following PEMFC steady-state operation—II. Influence of Ptz+ on membrane properties. J Electrochem Soc 154:B1115–B1120
13. Dubau L, Durst J, Maillard F, Guetaz L, Chatenet M, Andre J, Rossinot E (2011) Further insights into the durability of Pt3Co/C electrocatalysts: formation of "hollow" Pt nanoparticles induced by the Kirkendall effect. Electrochim Acta 56:10658–10667
14. Ferreira PJ, Ia OGJ, Shao-Horn Y, Morgan D, Makharia R, Kocha S, Gasteiger HA (2005) Instability of Pt/C electrocatalysts in proton-exchange membrane fuel cells. J Electrochem Soc 152:A2256–A2271
15. Shao-Horn Y, Sheng WC, Chen S, Ferreira PJ, Holby EF, Morgan D (2007) Instability of supported platinum nanoparticles in low-temperature fuel cells. Top Catal 46:285–305
16. Ramos-Sanchez G, Balbuena PB (2013) Interactions of platinum clusters with a graphite substrate. Phys Chem Chem Phys 15:11950–11959
17. Ma J, Habrioux A, Morais C, Lewera A, Vogel W, Verde-Gomez Y, Ramos-Sanchez G, Balbuena PB, Alonso-Vante N (2013) Spectroelectrochemical probing of the strong interaction between platinum nanoparticles and graphitic domains of carbon. ACS Catal 3:1940–1950
18. Franco AA, Tembely M (2007) Transient multiscale modeling of aging mechanisms in a PEFC cathode. J Electrochem Soc 154:B712–B723
19. Callejas-Tovar R, Balbuena PB (2011) Molecular dynamics simulations of surface oxide-water interactions on Pt(111) and Pt/PtCo/Pt3Co(111). Phys Chem Chem Phys 13:20461–20470
20. McMillan N, Lele T, Snively C, Lauterbach J (2005) Subsurface oxygen formation on Pt(100): experiments and modeling. Catal Today 105:244–253
21. Walker AV, Klotzer B, King DA (2000) The formation of subsurface oxygen on Pt{110} (1 × 2) from molecular-beam-generated O-2 (1)Delta(g). J Chem Phys 112:8631–8636
22. Over H, Seitsonen AP (2002) Oxidation of metal surfaces. Science 297:2003–2005
23. Gu Z, Balbuena PB (2007) Absorption of atomic oxygen into subsurfaces of Pt(100) and Pt (111): density functional theory study. J Phys Chem C 111:9877–9883
24. Gu Z, Balbuena PB (2007) Chemical environment effects on the atomic oxygen absorption into Pt(111) subsurfaces. J Phys Chem C 111:17388–17396
25. Gu Z, Balbuena PB (2008) Atomic oxygen absorption into Pt-based alloy subsurfaces. J Phys Chem C 112:5057–5065
26. Allen MP, Tildesley DJ (1990) Computer simulation of liquids. Oxford University Press, Oxford
27. Koper MTM, Jansen APJ, vanSanten RA, Lukkien JJ, Hilbers PAJ (1998) Monte Carlo simulations of a simple model for the electrocatalytic CO oxidation on platinum. J Chem Phys 109:6051–6062
28. Koper MTM, Lukkien JJ, Jansen APJ, Van Santen RA (1999) Lattice gas model for CO electrooxidation on Pt-Ru bimetallic surfaces. J Phys Chem B 103:5522–5529
29. Lukkien J, Segers JPL, Hilbers PAJ, Gelten RJ, Jansen APJ (1998) Efficient Monte Carlo methods for the simulation of catalytic surface reactions. Phys Rev E 58:2598–2610
30. Mainardi DS, Calvo SR, Jansen APJ, Lukkien JJ, Balbuena PB (2003) Dynamic Monte Carlo simulations of O$_2$ adsorption and reaction on Pt(111). Chem Phys Lett 382:553–560
31. Van Gelten RJ, Jansen APJ, Van Santen RA, Lukkien JJ, Segers JPL, Hilbergs PAJ (1998) Monte Carlo simulations of a surface reaction model showing spatio-temporal pattern formations and oscillations. J Chem Phys 108:5921–5934
32. Voter AF (2007) Introduction To The kinetic Monte Carlo method, vol 235
33. Callejas-Tovar R, Diaz CA, Hoz JMMdl, Balbuena PB (2013) Dealloying of platinum-based alloy catalysts: kinetic Monte Carlo simulations. Electrochim Acta 101:326–333
34. Kirkendall E, Thomassen L, Uethegrove C (1939) Rates of diffusion copper and zinc in alpha brass. Trans Am Inst Mining Metall Eng 133:186–203
35. Erlebacher J, Aziz MJ, Karma A, Dimitrov N, Sieradzki K (2001) Evolution of nanoporosity in dealloying. Nature 410:450–453

36. Holby EF, Greeley J, Morgan D (2012) Thermodynamics and hysteresis of oxide formation and removal on platinum (111) surfaces. J Phys Chem C 116:9942–9946
37. Wakisaka M, Asizawa S, Uchida H, Watanabe M (2010) In situ STM observation of morphological changes of the Pt(111) electrode surface during potential cycling in 10 mM HF solution. Phys Chem Chem Phys 12:4184–4190
38. Hirunsit P, Balbuena PB (2009) Surface atomic distribution and water adsorption on PtCo alloys. Surf Sci 603:911–919
39. Pokhmurskii V, Korniy S, Kopylets V (2011) Computer simulation of binary platinum-cobalt nanoclusters interaction with oxygen. J Cluster Sci 22:449–458
40. Yang ZX, Yu XH, Ma DW (2009) Adsorption and diffusion of oxygen atom on Pt3Ni(111) surface with Pt-skin. Acta Phys Chim Sin 25:2329–2335
41. Leisenberger FP, Koller G, Sock M, Surnev S, Ramsey MG, Netzer FP, Klotzer B, Hayek K (2000) Surface and subsurface oxygen on Pd(111). Surf Sci 445:380–393
42. Hammer B, Hansen LB, Norskov JK (1999) Improved adsorption energetics within density-functional theory using revised Perdew-Burke-Ernzerhof functionals. Phys Rev B 59:7413–7421
43. Hirunsit P, Balbuena PB (2009) Effects of water and electric field on atomic oxygen adsorption on PtCo alloys. Surf Sci 603:3239–3248
44. Tang W, Henkelman G (2009) Charge redistribution in core-shell nanoparticles to promote oxygen reduction. J Chem Phys 130
45. Norskov JK, Rossmeisl J, Logadottir A, Lindqvist L, Kitchin JR, Bligaard T, Jonsson H (2004) Origin of the overpotential for oxygen reduction at a fuel-cell cathode. J Phys Chem B 108:17886–17892

Molecular Dynamics Simulations
of Electrochemical Energy Storage Devices

Dario Marrocchelli, Céline Merlet and Mathieu Salanne

Abstract Many modelling problems in materials science involve finite temperature simulations with a realistic representation of the interatomic interactions. These problems often necessitate the use of large simulation cells or long run times, which puts them outside the range of direct first-principles simulation. This is particularly the case for energy storage systems, such as batteries or supercapacitors. For battery materials, it is possible to introduce polarizable potentials for the interactions, in which additional degrees of freedom provide a representation of the response of the electronic structure of the ions to their changing coordination environments. Such force field can be built on a purely first principles basis. Here we discuss the example of a Li-ion conductor, and we show how the long molecular dynamics simulations are useful for characterizing accurately the conduction mechanism. In particular, strong cooperative effects are observed, which impact strongly the electrical conductivity of the material. In the case of supercapacitors, the full electrochemical device can be modelled. However, this leads to very large simulation cell, which does not allow using polarizable force fields for the electrolytes. For the electrodes, fluctuating charges are used in order to maintain a constant electric potential as in electrochemical experiments. These simulations have allowed for a deep understanding of the charging mechanism of supercapacitors. In particular, the desolvation of the ions inside the porous carbon electrodes and the fast dynamic of charging can now be understood at the molecular scale.

D. Marrocchelli
Department of Nuclear Science and Engineering, Massachusetts Institute of Technology,
Cambridge, MA, USA

C. Merlet
Department of Chemistry, University of Cambridge, Lensfield Road,
Cambridge CB2 1EW, UK

M. Salanne (✉)
Sorbonne University, UPMC Univ Paris 06, UMR 8234, PHENIX, 75005 Paris, France
e-mail: mathieu.salanne@upmc.fr

M. Salanne
Réseau sur le Stockage Electrochimique de l'Energie (RS2E), FR CNRS 3459,
80039 Amiens Cedex, France

© Springer-Verlag London 2016   61
A.A. Franco et al. (eds.), *Physical Multiscale Modeling and Numerical
Simulation of Electrochemical Devices for Energy Conversion and Storage*,
Green Energy and Technology, DOI 10.1007/978-1-4471-5677-2_3

1 Introduction

How much energy can I store in a device? How fast can it be charged? These two questions are at the heart of today's research on electricity storage and are related to the dynamics of charge transport in these devices. Here we will focus on the two families of electrochemical energy storage devices that are currently most intensively studied, namely Li-ion batteries and supercapacitors. Li-ion batteries and supercapacitors differ significantly by the mechanism which allows them to store the electricity: In Li-ion batteries the mechanism is based on redox reactions occurring in the bulk electrodes, accompanied by Li^+ insertion/extraction, while supercapacitors accumulate charge at the surface of the electrodes by reversible adsorption of the ions of the electrolyte. This implies that different challenges have to be solved for modelling them accurately, whatever simulation technique is used.

Li-ion batteries store much greater amounts of energy per unit weight or volume compared to other rechargeable battery systems [1]. For this reason, they are extensively used in portable electronics and are now being considered for transportation applications, such as hybrid and electric vehicles, as well as for the efficient storage and utilization of intermittent renewable energies, like solar and wind (the so-called peak shaving) [2]. However, significant improvements are needed in the cost of these devices, their safety and cycle life. In addition, the energy and power densities have to be further enhanced to address the range anxiety of electric vehicles, i.e. the fact that most electric vehicles have relatively low driving ranges, usually below 150 km, which is unappealing to customers.

The modelling of Li-ion batteries has so far mainly involved ab initio approaches based on the density functional theory (DFT), which have a powerful predictive character in terms of structure and thermodynamics of the systems [3]. The drawback is its computational cost which remains prohibitive for studying the dynamics of the systems, which is essential for predicting whether an electrode material will be active or not, or screening potentially performant electrolytes. Such predictions are highly desirable because the experimental methods which allow for an unambiguous decoupling of the ionic and the electronic conductivity of the materials are very scarce [4], and because very often the values extracted can be very different from the real ones for the pure phases, due to the presence of defects, impurities, etc. [5]. The estimation of transport properties from DFT almost always relies on the determination of the energy landscape experienced by Li^+ ions, which is a static picture. As a consequence, the extracted diffusion coefficients often exceed the measured ones, sometimes by several orders of magnitudes. Ceder et al. [6] proposed an elegant interpretation of this discrepancy based on the fact that a vast majority of materials, including $LiFePO_4$, present some privileged 1D diffusion channels which can be blocked in macroscopic samples by immobile point defects that are not included in the relatively small simulation cells involved in the calculations. Nevertheless, there is currently no ab initio approach which is able to predict the diffusion coefficients and the electrical conductivity of the materials.

Proposing new simulation strategies, where the dynamics of the system is explicitly taken into account, is therefore mandatory for understanding completely the transport of Li^+ ions in battery materials.

The study of supercapacitors also represents a challenge as a result of the complexity and multi-scale nature of the materials used as electrodes and the requirement for an understanding of the evolution of the interface with the applied potential. In supercapacitors, the energy is stored at the electrode/electrolyte interface through reversible ion adsorption. Among all the potential electrode materials, carbon is the most commonly used as a result of its low cost, its relatively good electrical conductivity and the fact that it can be synthesized in different ways giving rise to a large variety of morphologies. Specifically, since the energy is stored through a surface phenomenon, porous carbon materials with large surface areas lead to higher densities. Nevertheless, the surface area is not the only material property that matters. In 2006, it was demonstrated that ions from the electrolyte can enter pores of sub-nanometer sizes leading to a huge increase of capacitance (+100 % volumetric capacitance compared to mesoporous carbons) [7]. This finding has generated a great deal of technological activity to refine potential devices and fundamental research to examine the underlying molecular phenomena.

To fully understand the interplay between the electrode structure, the electrolyte nature and the resulting electrochemical properties, experimental and theoretical methods should be able to deal with *complex porous materials* and to probe *local processes*. On the experimental point of view, techniques suitable for the detailed study of supercapacitors are still rare and the establishment of in situ methods, able to probe the local behaviour of the interface at various potentials, represents substantial breakthroughs. Very recently, in situ Nuclear Magnetic Resonance (NMR) [8, 9] and Electrochemical Quartz Crystal Microbalance (EQCM) [10–12] have been developed to allow for the characterisation of the structural and dynamical properties of the electrode/electrolyte interface. While these techniques are very powerful, running these experiments is extremely complicated, the interpretation of the resulting data is not straightforward and theoretical studies are usually necessary to help/validate the data analysis.

For studying both Li-ion batteries and supercapacitors, Molecular Dynamics (MD), a simulation technique which yields the trajectory of the atoms, appears as the most appropriate tool for studying the dynamics of the systems. Unfortunately, although ab initio MD studies have started to appear recently in the case of Li-ion batteries, [13–15] they are limited to small system sizes and short simulation times (100 ps), which hinders the calculation of accurate diffusion coefficients or electrical conductivities as well as any effect of microstructure (grain boundaries, dislocations, etc.). On the contrary, classical MD gives access to larger simulation cells and longer trajectories, and it has been proven useful for the study of transport properties in many scientific fields [16–20]. Its application to electricity storage devices has been relatively hindered by the lack of correct models to represent the systems at play. Indeed, the predictive power of classical molecular simulations mainly depends on how accurately the interactions between atoms/molecules are treated; in classical MD these interactions are described by an analytical force field derived from a model. Depending on the system of interest and the properties that have to be sampled, the level of sophistication of the model has to be adapted. The first section of this chapter will therefore be dedicated to the description of the models that we have recently proposed for the study of electrochemical devices. Then we will discuss separately the two types of devices, Li-ion batteries and supercapacitors. In the last section we will discuss the perspectives which could be opened by extending further the models used in the simulations.

2 Molecular Dynamics

2.1 Principle

MD is nowadays one of the most used molecular simulation techniques. Its principle is simple: Given the coordinates and velocities of a set of atoms at a given time t, their new values at a small amount of time later $t + \Delta t$ are numerically calculated by using Newton's equation of motion; Δt is called the timestep. This is done several thousands or millions of times in order to generate a representative trajectory of the system, which can be done in practice by many different algorithms. Depending on the properties which are targeted, the *NVE* (i.e. constant number of atoms N, volume V and energy E), *NVT* (constant N, V and temperature T) or *NPT* (constant N, pressure P and T) ensembles are generally sampled.

The most important ingredient of a MD simulation is the interaction potential, also named force field, from which the forces which act on the atoms are derived at each step. Hundreds of models exist [21–23], which have various levels of complexity, but only a few of them are routinely used and included in standard simulation packages. Electrochemical systems are intrinsically difficult to model, and we must resort to advanced models, which are described in the following.

2.2 All-Atom Force Fields

A realistic force field must account not only for the classical electrostatic interaction, but also for three interactions arising from the quantum nature of electrons: The exchange–repulsion (van der Waals repulsion) is a consequence of the Pauli principle, while the dispersion (van der Waals attraction) arises from correlated fluctuations of the electrons. Lastly, the polarization (induction) term reflects the distortion of the electron density in response to electric fields.

In oxides, all these terms are included by using the following analytic form of the interaction potential between each pair of atoms:

$$U_{ij}(r_{ij}) = \frac{q_i q_j}{r_{ij}} + \frac{q_i r_{ij} \cdot \mu_j}{r_{ij}^3} f_4^{ij}(r_{ij}) - \frac{\mu_i \cdot r_{ij} q_j}{r_{ij}^3} f_4^{ji}(r_{ij})$$

$$+ \frac{\mu_i \cdot \mu_j}{r_{ij}^3} - \frac{3(r_{ij} \cdot \mu_i)(r_{ij} \cdot \mu_j)}{r_{ij}^5}$$

$$+ B_{ij} \exp(-\alpha_{ij} r_{ij}) - \frac{C_6^{ij}}{r_{ij}^6} f_6^{ij}(r_{ij}) - \frac{C_8^{ij}}{r_{ij}^8} f_8^{ij}(r_{ij}) \qquad (1)$$

where q_i and μ_i are the charge and dipole moment of particle i, respectively, and f_n^{ij} the damping function for short-range correction of interactions between charge and dipole and dispersion interactions [24]:

$$f_n^{ij}(r^{ij}) = 1 - c_n^{ij} e^{-b_n^{ij} r^{ij}} \sum_{k=0}^{n} \frac{(b_n^{ij} r^{ij})^k}{k!}. \qquad (2)$$

$\{B_{ij}, \alpha_{ij}, C_6^{ij}, C_8^{ij}, b_n^{ij}, c_n^{ij}\}$ are a set of parameters. We have developed a methodology which allows us to determine them from first-principles calculations, i.e. without including any experimental information [21].

The dipole moments induced on each ion are a function of the polarizability and electric field on the ion caused by charges and dipole moments of all the other ions. They are determined at every time step self-consistently using the conjugate gradient method by minimization of the total energy. The charge-charge, charge-dipole, and dipole-dipole contributions to the potential energy and forces on each ion are evaluated under the periodic boundary condition by using the Ewald summation technique [25].

2.3 Modelling Metallic Electrodes at Constant Potential

The electrodes are metallic materials, in which the potential is kept constant during the experiments. In our simulations, they are modelled following a method developed by Reed et al. [26], following a proposal by Siepmann and Sprik [27], which consists in determining the charge on each electrode atom at each MD step by requiring that the potential on this atom is constant and equal to a specified value. This condition therefore means that the electrode polarizes in the same way as a metal, where the potential inside the metal is a constant. The results, in particular dynamical properties, will be very different from results obtained by a simpler and faster but less accurate technique, which consist in assigning constant charges to the electrode atoms [28]. In the constant potential method, the electrode atom charges, represented by Gaussian distributions centred on the carbon atoms, are obtained by minimizing the expression:

$$U^{\text{electrode}} = \sum_i q_i(t) \left[\frac{\Psi_i(\{q_j(t)\})}{2} + \frac{q_i(t)k}{\sqrt{2\pi}} - \Psi^0 \right],$$
(3)

where q_i is the charge on electrode atom i, κ is the width of the Gaussian distributions ($\kappa = 0.5055$ Å in this work), $\Psi_i(\{q_j\})$ is the potential at position i due to all the ions in the electrolyte and the other electrode atoms, the second term originates from the interaction of the Gaussian distribution with itself and Ψ^0 is the externally applied potential. Instead of minimizing this expression using a conjugate gradient method only, we use a two-step procedure where the first step is a prediction of the charges by a polynomial expansion, using the previous steps [29, 30], and the second step is a conjugate gradient minimization. This allows a speed-up by a factor of 2 compared to conjugate gradient minimization only.

2.4 Coarse-Grained Force Fields

In the simulations of supercapacitors, all the components of the devices have to be included in the simulation cells. In addition, the electrolyte ions are large molecular species and the two electrodes are held at constant potential, which increases greatly the simulation cost. Using an all-atom, polarizable force field for the electrolyte ions is therefore impossible, even with the current supercomputers. It is possible to overcome this difficulty by using a simplified model, in which the atomic details of the molecules are averaged out. This approach is named coarse-graining, and the interaction between two grains is then represented by the Lennard-Jones analytical form:

$$U_{ij}(r_{ij}) = 4\epsilon_{ij} \left[\left(\frac{\sigma_{ij}}{r_{ij}} \right)^{12} - \left(\frac{\sigma_{ij}}{r_{ij}} \right)^6 \right] + \frac{q_i q_j}{r_{ij}}$$
(4)

For the 1-butyl-3-methylimidazolium hexafluorophosphate (BMI-PF$_6$) ionic liquids, each cation is represented by 3 sites (instead of 19 atoms) and each anion by one site only (instead of 7 atoms). The number of parameters is much reduced compared to all-atom force fields, and we use the values proposed by Roy and Maroncelli. Their force field accurately predicts a large number of bulk properties of the pure ionic liquid [31] such as molar volume, isothermal compressibility, viscosity and diffusion. For systems in which an acetonitrile solvent is added, we have used the model developed by Edwards et al. [32] in which three sites are used to describe the whole molecule.

3　Li-Ion Batteries

The current limitations of Li-ion batteries are mainly linked to the nature of employed electrode and electrolyte materials. Spurred by these limitations, the search of *new materials* for lithium-ion batteries with enhanced performance has received much attention in the recent years [33]. This has resulted in the discovery of many new classes of promising materials for cathode applications, such as phosphates (e.g. $LiFePO_4$) [34–36], pyrophosphates (e.g. $Li_2FeP_2O_7$) [37, 38] or more recently silicates (e.g. Li_2FeSiO_4) [39, 40] and sulphates ($LiFeSO_4F$) [5, 41, 42]. New solid-state electrolyte materials have also been found; indeed, Kamaya et al. have recently reported on a new superionic conductor, $Li_{10}GeP_2S_{12}$, that presents a conductivity as high as 12 ms cm^{-1} at room temperature [43]. This represents the highest conductivity achieved in a solid electrolyte, exceeding even those of liquid organic electrolytes.

Despite the exciting and promising recent progress in finding new materials, many challenges still remain. One key challenge is to improve the Li-ion conductivity as this property strongly affects the battery performance (e.g. higher Li conductivity allows faster charging). Because of its importance, a significant research effort has been made over the past few years to understand the factors affecting Li-ion conductivity and to find materials with higher conductivity. To this end, modelling techniques have been intensively used in the hope to accelerate and advance the development of more conducting materials for battery applications [13, 15, 44–46]. The customary approach is to perform *static* calculations (based either on DFT [47–51] or empirical interatomic potentials) to then evaluate migration energies for the Li-ion diffusion process [3, 13, 45, 46, 52]. These migration energies can then be related to the diffusion coefficient via the well-known equation, $D = a^2 v^* \exp(-E_{act}/k_B T)$, (where a is the Li^+ ion hopping length, v^* is the hop attempt frequency and E_{act} is the activation barrier). This oversimplified approach, however, can lead to significantly overestimated diffusion coefficients [53]. Also the use of DFT calculations severely limits the size of the systems that can be studied, thus hindering, e.g., the study of the effects of micro-structure (grain boundaries, interfaces, etc.) on this property. In principle, an alternative approach would be the use of MD simulations that allow the diffusion process to be studied directly from the displacements of the Li-ion. These simulations, however, rely on the accuracy of the employed interatomic potential and some of the empirical potentials available in the literature have been shown to fail in certain cases [46, 54].

In the remainder of this section, we will review some of our recent work [55, 56] on $LiMgSO_4F$, a structural analogue of $LiFeSO_4F$. This recently discovered material has a very high voltage (3.6–3.9 V vs. Li), suppresses the need for nano-sizing (thanks to its higher Li-ion conduction) and has a significant cost advantage compared to other materials [41]. These properties make it a promising candidate for the cathode material in higher volume applications, such as electric vehicles. Previous computational work, however, failed at predicting its diffusion coefficient, with both DFT [3] and empirical calculations [46] yielding much higher values than those obtained experimentally. Similar issues were encountered with similar materials, such as $LiFePO_4$. By performing reliable MD simulations using accurate inter-ionic potentials, parameterized with respect to *first-principles* calculations, we showed that this material presents a cooperative conduction mechanism and that this explains the discrepancy between the calculated and measured diffusion coefficients.

At this point we would like to note that this section is not intended, by any means, to be a comprehensive overview of MD simulations of Li-ion batteries (a subject worth a book on its own!), but rather a simple example of the challenges and opportunities in this field. We therefore refer an interested reader to other references, such as the recent special issue of *Modeling and Simulations in Materials Science and Engineering* [57] on this subject.

3.1 A Polarizable Force Field Based on First-Principles Calculations

As mentioned in the previous sections, the predictive power of classical molecular simulations mainly depends on how accurately the interactions between atoms/molecules are treated. In the work pioneered by Madden et al. [21], two key factors were found to be of paramount importance for obtaining an accurate interaction between atoms/molecules. The first factor is the use of potentials that are physically well justified, such as the one reported in Eq. (1). Indeed, in the case of many ionic systems, it has been found that including in the potential an energy term related to the ionic polarizability of the anions and, to a lesser extent, of the cations is necessary, and often sufficient, to yield an accurate description of the interactions between ions [16–20, 53, 58–62].

The second important factor has to do with the way potentials are parameterized. In most cases found in the literature, interionic potentials are parameterized by fitting them to a few known experimental quantities, such as lattice parameters and elastic constants [45, 46, 52]. This fit is sometimes performed manually, though more rigorous, systematic approaches are available (e.g. the well-known GULP code has an automated fitting routine). This approach has quite a few drawbacks: The number of data for the fitting is usually small (often there are more parameters in the potential that data points to fit!); also the potential is, by definition, empirical, which means that only systems for which experimental data are already available can be studied. A much better approach is to use DFT calculations to generate a data set that can then be used to fit the potential. This approach solves the drawbacks mentioned above, as one can easily generate a wealth of data points ($\sim 10,000$) for the potential fit and the potential is of *first-principles* accuracy, since no empirical information was used during the fit.

We applied this approach to fit an interionic potential for $LiMgSO_4F$. We ran a series of DFT calculations on super cells containing 128 atoms. From these simulations, we extracted the forces acting on each atoms and atomic dipoles (calculated from a Wannier analysis) [63]. Both forces and dipoles formed a data-set of more than 3000 points against which we fitted the parameters of our potential.

The potential was validated by checking its ability to reproduce the structural properties of this material. This approach has been used in several oxide [16–20, 60, 61] and halide systems, where these potentials have been shown to be very accurate and highly predictive. We note here that, since no experimental information was used in the fit, the fact that this potential can reproduce these properties represents a very good assessment of its reliability.

The lattice parameters of $LiMgSO_4F$ in the tavorite structure were calculated via a structure minimization. These were found to agree very well with the experimental values, the maximum deviation being less than 1 %. Such a close agreement illustrates the accuracy of the developed potential. Bond distances were also extracted from short MD runs at room temperature and were found to agree well with structural data from diffraction experiments. Finally, the Li-ion positions (obtained from a MD run and reported in Fig. 2 of Ref. [55]) also agree with the structural data. In conclusion, our potential can successfully reproduce the structural properties of this material, yielding an agreement comparable to that obtained with DFT calculations (though at a fraction of the computational cost).

3.2 Conduction Mechanism in Stoichiometric LiMgSO₄F

With our computationally efficient interionic potential we were able to simulate a $6 \times 6 \times 4$ supercell, containing 288 $LiMgSO_4F$ unit cells (2304 atoms) for as long as 8 ns, with little computational effort.[1] This represents a significant improvement over DFT methods, which can at most simulate ~ 100 atoms for a few tens of ps. In our simulations, we observe clear Li^+ diffusion for temperatures greater than 1100 K. Figure 1a shows the coordinates of two neighboring Li^+ ions during the course of a 4 ns simulation at 1200 K. These ions clearly diffuse by a series of consecutive hops. Interestingly, these hops appear to be very *correlated*, as shown by the fact that both ions hop at exactly the same time, in most cases. From the run at 1200 K, Mean Squared Displacements (MSD) were extracted and these are reported in Fig. 1b. Here we can clearly observe three regimes in the MSD curve: ballistic, caging and diffusive. The first regime (for very short times) corresponds to the unhindered motion of the Li^+ ions, i.e. a ballistic motion. Very soon, however, the ions bounce back because of the presence of neighboring F and O ions; this is the so called caging regime, where the ions are trapped at their sites and can only vibrate around their equilibrium position. Eventually, after a certain time, the ions can break free and jump to a neighboring position. This is the diffusive regime.

---

[1]These simulations were run on 64×2.66 GHz Intel cores for about 3 days.

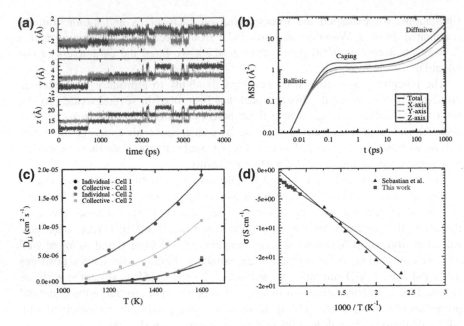

Fig. 1 **a** Coordinates of two Li$^+$ ions during a simulation performed at 1200 K. **b** Mean squared displacement (total and X, Y, Z components) of the Li$^+$ ions at 1200 K. **c** Individual and collective diffusion coefficients for two cell parameters. **d** Comparison between the experimental (*blue squares*) and experimental conductivity (*black triangles*). The lines are Arrhenius fits. Adapted with permission from Ref. [55]. Copyright (2012) American Chemical Society (Color figure online)

The duration of the caging regime depends on the temperature; in the case reported in Fig. 1b, it takes them ∼100 ps to break free. We emphasize that such long timescales are basically unaccessible with DFT calculations, which again points to the importance of performing MD simulations with interionic potentials.

As mentioned above, the diffusion mechanism of this material has a cooperative character, as shown by the fact that, most of the times, neighboring ions hop at exactly the same time (see Fig. 1a). For this reason, here we briefly summarize how to extract conductivities from a system with a cooperative diffusion mechanism. Further information can be found in our previous work [55, 56] and in classical textbooks [64]. When performing MD simulations, the ionic conductivity of these materials is generally calculated from the diffusion coefficient, via the so-called Nernst-Einstein equation,

$$\sigma = \frac{\rho e^2}{k_B T} D, \qquad (5)$$

where ρ is the density, e the charge and D the diffusion coefficient of the conducting species (the Li$^+$ ions in this case). The diffusion coefficient is calculated from a MD run using the ionic mean squared displacements (such as the ones reported in Fig. 1b):

$$D = \lim_{t \to \infty} \frac{1}{6t} \langle |\delta r_i(t)|^2 \rangle. \tag{6}$$

This widely used approach has been employed in several computational studies of ionic conductors [17–19, 65–67]. However, this expression assumes that the motion of the conducting species is uncorrelated. When this approximation does not hold (as we have shown here), the conductivity should be calculated using the following equation:

$$\sigma = \frac{\rho e^2}{k_B T} D_{\text{eff}}, \tag{7}$$

where the term D_{eff} is now an effective diffusion coefficient that takes into account correlation effects and it is defined as:

$$D_{\text{eff}} = \frac{1}{N_{\text{Li}}} \lim_{t \to \infty} \frac{1}{6t} \langle | \sum_i^{N_{\text{Li}}} \delta r_i(t)|^2 \rangle. \tag{8}$$

Comparing the expression for D and D_{eff} one can understand right away why D_{eff} is seldom calculated. Indeed, for the calculation of D, one has to calculate the displacements of the individual ions at a certain time t and, then, take the average of these. On the other hand, when calculating D_{eff}, one does not average over the individual displacements but simply sums them up. For this reason, D_{eff} is usually a very noisy quantity that can only be calculated when the studied system has a sufficient number of conducting ions and the trajectory is long enough for a diffusive regime to be established. Again, we highlight the fact that MD simulations with interionic potentials are necessary to obtain reliable D_{eff} and that this would not be possible with DFT.

Finally, we note that the above expression for the conductivity is sometimes recast as:

$$\sigma = \frac{\rho e^2}{k_B T} D_{\text{eff}} = \frac{\rho_i e^2}{k_B T} HD, \tag{9}$$

where H is called the Haven ratio [68, 69] (defined as the ratio between D_{eff} and D) and is an indicator of how correlated the ionic motion is in a specific material.

In Fig. 1c we show the calculated D and D_{eff} as a function of temperature, for two slightly different cell parameters (see Ref. [55] for further details). We note that the difference between D and D_{eff} is significant, especially at low temperature, with D_{eff} being about an order of magnitude larger than D at 1200 K. This confirms the strongly cooperative character of the diffusion mechanism in this material; indeed, this corresponds to a Haven ratio of 10! Figure 1d shows the calculated ionic conductivity in an Arrhenius plot. This is compared to the experimental data of Sebastian et al. [70]. We note that the experimental data are limited to fairly low temperatures, while our calculations are at higher temperature. For this reason, we add Arrhenius fits to both data sets; the activation energies we obtain are in very good agreement (E_a = 0.84 and 0.94 eV for MD and experiments, respectively). Also, extrapolating the MD data to lower temperatures leads to a good agreement with the experimental data. At room temperature (the last black triangle in the graph) the difference between the MD and experimental value is a factor of 5, a significant improvement over the 3 and 5 orders of magnitude difference predicted by previous calculations [3, 46]. Our interionic potential can therefore reproduce the conducting properties of this system with very high accuracy.

Finally, we turn our attention to the atomistic causes of this cooperative conduction mechanism. To this end, we report an example of collective diffusion behaviour extracted from our simulation in Fig. 2. In this series of snapshots the

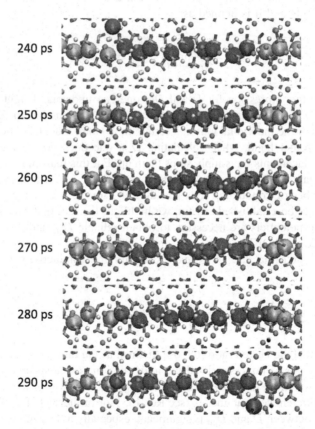

Fig. 2 Successive snapshots, taken along a [111] diffusion channel, highlighting the cooperative diffusion mechanism of Li$^+$ ions at 1200 K. The SO$_4{}^{2-}$ tetrahedra are represented by the S–O bonds, while the Mg and F atoms are, respectively, shown as *white* and *green spheres* (note that the atoms are not shown using their usual van der Waals radii in order to enhance the Li$^+$ ions positions). The Li$^+$ ions either appear as *pink, red, blue* and *purple spheres*. The red Li$^+$ ions jumps into the [111] channel, provoking a correlated displacement of all the blue ions, and finally the ejection of the purple one. Adapted with permission from Ref. [55]. Copyright (2012) American Chemical Society (Color figure online)

system is viewed along one of the [111] diffusion pathways. The SO_4^{2-} tetrahedra are represented by the S–O bonds, while the Mg and F atoms are, respectively, shown as white and green spheres (note that the atoms are not shown using their usual van der Waals radii in order to enhance the Li^+ ions positions). The Li^+ ions either appear as pink, red, blue and purple spheres: The pink ones do not experience any jump during this portion of trajectory. The red one, which is in a different [111] channel at $t = 240$ ps, suddenly jumps, at $t = 250$ ps, in the channel where the blue Li^+ atoms are. At that moment, the Li^+ ions represented in blue start to propagate by hops from the left to the right; at each snapshot in the time interval between 250 and 280 ps we observe a set of around 5 ions which are more closely packed, which shows the propagation of the excitation. The propagation stops at $t = 290$ ps, when the ion shown in purple is ejected in another [111] channel. In this example, if we consider the blue ions only, each of them has jumped by δr_{hop} while the overall charge was displaced by 9 times δr_{hop}, which highlights the difference between the individual and collective diffusion coefficients, and the non-validity of the Nernst-Einstein relationship for systems with correlated motion.

3.3 Effect of Li^+ Vacancies

In light of the findings reported above, a natural question to ask is: *What happens to the cooperative conduction mechanism observed in LiMgSO₄F (and related com-pounds), when we create Li vacancies in the system?* This question is of great technological interest because cathode materials undergo Li insertion/removal dur-ing charging/discharging of a battery. So one is interested in the conduction mechanism of these compounds for different Li contents. We attempted to simulate this by using a simplified approach, in which Li^+ vacancies are introduced in $Li_{1-y}MgSO_4F$ and charge neutrality is obtained by assigning a 3+ charge to some Mg ions. While we are aware of the limitations of this approach (which are discussed in depth in Ref. [56], we think this can still provide useful information on the role of Li vacancies without having to resort to more complicated potential forms that can describe charge transfer, such as the reaxFF potential. We also refer to the Perspectives section of this chapter, where we discuss future work in this direction.

In Fig. 3a we report the diffusion coefficients (both D and D_{eff}) as a function of the Li^+ content, at 1200 K. We span concentrations that run from fully stoichio-metric ($y = 0$) to almost fully de-lithiated ($y = 0.9$). The diffusion coefficients (black and red circles) undergo a maximum for $y = 0.5$. This maximum is probably caused by the fact that, as more Li vacancies are created, the mobility of the remaining Li-ions is increased at first (see also discussion below). However, after a critical concentration of Li vacancies, these will start interacting and ordering, thus reducing the Li-ion mobility. A similar phenomenon is observed in oxide-ion conductors [17–19, 71], though in this case the maximum is usually observed at lower vacancy concentrations (this points to a lower interaction energy between

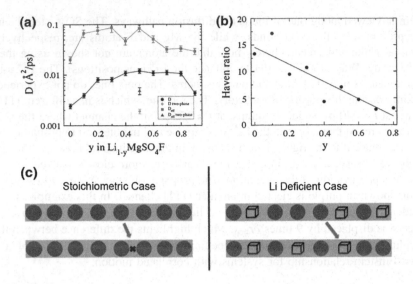

Fig. 3 **a** Calculated diffusion coefficients (D and D_{eff} as *black* and *red circles*, respectively) at 1200 K as a function of Li$^+$ content, for a system in which Mg^{2+} and Mg^{3+} are randomly distributed throughout the material. The *green* and *blue circles* are calculated from a two-phase material, see Ref. [56]. **b** Haven ratio as a function of y, calculated from the data in (**a**). **c** Schematic of the conduction mechanism in stoichiometric (*left*) and highly Li-deficient (*right*) LiMgSO$_4$F. The *black cross* and the *black cube* indicate a Li ion interstitial and vacancy, respectively. Adapted with permission from Ref. [56]. Copyright (2013) Institute of Physics (Color figure online)

vacancies in this system). We note here that these calculations are performed by randomly distributing the Mg^{2+} and Mg^{3+} throughout the material. This picture is consistent with a single-phase Li insertion/removal mechanism, i.e. a system in which the 2+ and 3+ Mg cations are randomly distributed. We also attempted to simulate a two-phase material and the resulting diffusion coefficients (blue and green dots) are significantly lower. We refer an interested reader to Ref. [56] for further details.

Figure 3b shows the Haven ratio (as defined above) calculated for different values of Li$^+$ content, y, at 1200 K. This quantity decreases significantly as more Li$^+$ vacancies are added, meaning that the motion of the Li-ions becomes less cooperative. In fact, this is quite intuitive, since, as more Li vacancies are created, the remaining Li-ions will be further away from each other and, therefore, the diffusion of one ion will not be much affected by the presence of another ion. Interestingly, this implies that, during charging/discharging of such cathode materials, the conduction mechanism changes from a strongly correlated one, for a fully lithiated material, to a weakly correlated one, for Li-deficient materials.

Activation energies were extracted from select values of Li$^+$ contents, namely $y = 0$, 0.5 and 0.8. These are reported in Table 1. The activation energy decreases significantly as y is increased: It is 0.84 eV for the stoichiometric material and 0.4 eV for $y = 0.8$. This behaviour is explained schematically in Fig. 3c. On the

Table 1 Activation energies for different Li$^+$ contents

y	E_a (eV)
0.0	0.84
0.5	0.41
0.8	0.40

left-hand side of the figure, we show two channels along the [111] direction for LiMgSO$_4$F. The Li-ion crystallographic sites are all filled, because there are no vacancies. For this reason, when an ion jumps from one channel to another one, it will create an interstitial defect in the new channel and leave behind a vacancy. The formation of this Frenkel pair will require a certain amount of energy, so that the activation energy of the ionic conduction will be the sum of a migration enthalpy (ΔH_m) plus a Frenkel defect formation energy ($E_{Frenkel}$). The situation is different for a highly Li-deficient material, as shown on the right hand side of Fig. 3c. In this case, since there are many vacancies in the system, a Li-ion can easily hop from one channel into a vacancy in another channel, without having to create a Frenkel pair. In this case, the activation energy of conduction will be given by the migration enthalpy alone. If we assume that the activation energy for $y = 0.8$ is equal to the migration enthalpy alone (ΔH_m), while, for $y = 0.0$, it corresponds to $\Delta H_m + E_{Frenkel}$, we can then estimate the energy of creation of a Frenkel pair. Here we obtain $E_{Frenkel} = 0.44$ eV and $\Delta H_m = 0.40$ eV. These numbers must be taken with caution, since the diffusion data for $y = 0.8$ is quite noisy, due to the fact that very few Li-ions (only 48!) are present in this system. However, it is interesting to see that the value obtained for the migration enthalpy, $\Delta H_m = 0.40$ eV, is in closer agreement with previous static calculations by Tripathi et al. [46] ~ 0.4 eV and Mueller et al. [3], who extracted a value of 0.21 eV. This also explains why certain systems are much better ionic conductors than others, even though the Li-ion migration enthalpies are comparable. Take the example of the recent superionic conductor, Li$_{10}$GeP$_2$S$_{12}$. This material has a very low Li-ion migration enthalpy (~ 0.21–0.24 eV [13, 43]) and several empty crystallographic sites, so that the activation energy of conduction does correspond to the migration enthalpy alone.

3.4 On the Importance of Finite-Size Effects

In summary, in our recent work, we have derived an accurate interionic potential for LiMgSO$_4$F from *first-principles*. These were then used to perform extensive MD simulations to study the conduction mechanism of this material. We found that our potential can reproduce very accurately both the experimental structural and conduction data for this material, a significant improvement over previous models. Also, the conduction in this material is highly cooperative, especially when no Li$^+$ vacancies are present.

We note here that the cooperative character of the Li diffusion in this material has also important technical consequences. Firstly, it appears clearly that the hypothesis that the Li^+ ions are not interacting is not valid in this system, preventing the use of Eq. (5) (at least with the commonly used parameters). Secondly, we found out that special care must be taken when choosing the simulation box size. Indeed the excitation shown in Fig. 2 involves 9 Li^+ ions in the channel. This implies that simulation cells in which the channels contain at least twice this number of ions have to be used to avoid the spurious interaction of some Li^+ ions with their periodic images. Indeed, when relatively small systems (96 $LiMgSO_4F$ units) were simulated, the calculated mean squared displacements were noisy and the extraction of the diffusion coefficient was very difficult, even when trajectories of several tens of nanoseconds were used. We therefore performed calculations on systems sizes ranging from 96 to 1600 $LiMgSO_4F$ units, corresponding to 768–12,000 atoms. We found that the collective diffusion coefficients converge as the system size is increased and are virtually system size independent when 6144 (768 $LiMgSO_4F$ units) atoms, or more, are simulated. In our previous work, we have reported the results obtained from a supercell containing 288 $LiMgSO_4F$ units, corresponding to 2304 atoms (the 288 Li^+ ions are shared between 12 channels of 24 ions each). The diffusion coefficients extracted from this were very close to the converged values obtained from bigger systems, and this supercell allowed us to perform long simulation and thus to provide sufficient statistic at relatively low temperatures (1100–1300 K).

4 Supercapacitors

4.1 Increase of the Capacitance in Nanoporous Carbons

The finding that the use of microporous carbons instead of mesoporous as electrode materials for capacitors leads to a large increase of capacitance [7, 72–74] generated a great deal of activity in the modelling community in an effort to try to rationalize this discovery. Traditional theories, developed to analyse the case of a dilute electrolyte at a planar interface, are not applicable to supercapacitors in which highly concentrated electrolytes share an interface with complex electrodes. Recently, some analytical theories have been developed to include ion sizes and electrode polarization [75, 76]. These theories allow for the qualitative explanation of a number of phenomena and possible interpretations of the complex voltage–capacitance curves obtained experimentally [77–81]. Nevertheless, mean field theories still miss the inclusion of key factors such as correlation between molecules/ions and a realistic representation of the electrode structure with its heterogeneity and connectivity characteristics.

MD studies appear as a key tool in that respect as they allow for the inclusion of these effects assuming that a sufficiently accurate force field is chosen to represent both the electrode and the electrolyte. Most of the MD simulations relevant to supercapacitors are currently done with ideal electrode structures such as planar electrodes [26, 82–94], carbon nano-onions [95] and carbon nanotubes [96–99]. In the case of planar electrodes, the liquid adopts a layered structure at the interface as a consequence of the packing of the ions. This finding, first reported by Heyes and Clarke [100], has now been confirmed in a large number of simulation studies and observed experimentally as well using for example Atomic Force Microscopy [101–103] and Surface Force Apparatus [104] techniques.

This layered structure is in fact associated with an overscreening effect also well described in the literature [76, 83–85, 105]. The charge in the first layer of ions adsorbed at the electrode is higher than the electrode charge leading to the existence of a residual charge which is again overcompensated in the second layer of ions. This effect tends to extend the layering in the liquid side of the interface and reduce the charge storage efficiency.

In our MD simulations, we studied a supercapacitor consisting of either ionic liquids or acetonitrile-based electrolytes and carbide-derived carbons (CDCs) electrodes [106, 107]. CDCs are nanoporous carbons with well-defined pore size distribution [108, 109], for which a strong increase of the capacitance with decreasing pore size was shown experimentally for various electrolytes [7, 73, 110, 111]. In our simulations, we have used atomic structures which were generated by Palmer et al. using quenched MD [112] and were shown to correspond to CDCs synthesized from crystalline titanium carbide using different chlorination temperatures: They have similar pore size distributions, accessible surface and porous volumes as the experimental ones. In order to understand the origin of the increase of the capacitance in CDCs compared to mesoporous carbon, we also performed simulations with simpler, planar graphite electrodes. Typical snapshots of the simulation cells are shown in Fig. 4 in the case of an electrolyte composed of 1-butyl-3-methylimidazolium cations and hexafluorophosphate anions dissolved in acetonitrile. All our MD simulations were performed at fixed electrode potential using the technique explained in the methodology section.

When we study the local organization of the liquid inside the pores, special care must be taken in correctly defining the electrode surface. Following the experimental procedure, we define it as the surface accessible to an argon atom probe; the ionic density profiles are then calculated with respect to the normal to the local surface [113]. The first point we need to address is whether the ions are adsorbed on the surface of porous carbon in the same manner as on planar graphite electrodes. From Fig. 5, which reports the density profiles in both cases, it is immediately seen that the situation is very different: In the case of the porous electrodes, both cations and anions are allowed to approach the surface more closely by ≈ 0.7 Å. In a parallel-plate capacitor, the capacitance varies as the inverse of the distance between the two charged planes, suggesting that this shorter carbon atom–ion distance is partly at the origin of the capacitance increase in nanoporous electrodes. The fact that the ionic density profiles, shown in Fig. 5, seem smaller for the porous carbons than for the graphite is rather counterintuitive given that the capacitances behave in the opposite way. The integration of the ionic density profiles shown in

[BMI][PF₆] in ACN (1.5M)
Graphite electrodes

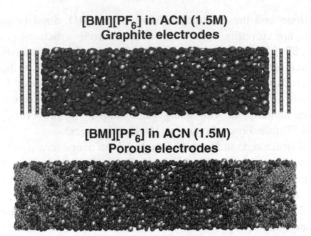

[BMI][PF₆] in ACN (1.5M)
Porous electrodes

Fig. 4 Typical snapshots of the simulated supercapacitors. Carbon–carbon bonds in the electrode are represented by turquoise sticks, while hexafluorophosphate anions, 1-butyl-3-methylimidazolium cations and acetonitrile molecules are, respectively, represented by *green*, *red* and *blue spheres* (Color figure online)

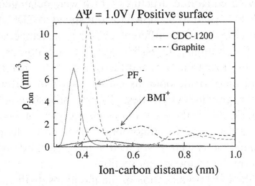

Fig. 5 Ionic density profiles normal to the electrode surface for graphite and CDC material. An electric potential of 1 V is applied between the electrodes and the electrolyte is a solution of BMI-PF₆ in acetonitrile. The distances are given with respect to the surface accessible to an argon atom probe, with the origin set to the position of the carbon atoms [106]

Fig. 5 provides the number of ions adsorbed at the surface of the electrode. For both types of ions this number is much smaller for the porous carbon than for the graphite electrodes. This result may seem rather counterintuitive given that the capacitance behaves in the opposite way, and suggests that a different charging mechanism is at play.

Since the liquid structure at the interface with a planar electrode is characterized by important overscreening effects, we have calculated the total surface charge accumulated across the liquid side of the interface. We have shown that for an applied potential difference of 1 V between the electrodes, the charge in the first adsorbed layer on a graphite electrode reaches a value which is two or three times higher than the charge of the electrode itself [106, 114]. This behaviour, due to the overscreening effect mentioned above and arising from ionic correlations, is observed for both the negative and positive graphite electrodes: The polarization of the first layer is coupled to that of the next layers. As a result, only a fraction of the adsorbed ions are effectively used in the electricity storage process. On the contrary, for porous electrodes, there is only one adsorbed layer, and the charging mechanism involves the exchange of ions with the bulk liquid. Consequently, the total charge in this first layer balances exactly that of the electrode, which results in a much better efficiency. Because the attraction of the ions in the first layer to the carbon surface is not balanced by that of a well-organized second layer, these ions approach the surface more closely than in the planar case. This original mechanism accounts for the larger charge stored inside nanoporous electrodes and reveals its microscopic origins [106].

4.2 Effect of the Local Structure

Although nanoporous structures generally exhibit high storage capabilities, their capacitance can vary a lot when passing from one carbon to another (in experiments, this is done, for example in the family of CDCs by changing the synthesis temperature), even when the materials have similar pore size distributions [106]. In fact, they show different local features such as small graphitic domains. This indicates that the local structure can affect the capacitance value. In order to characterize it, we have used the simple concept of coordination numbers. As depicted in Fig. 6, by defining a spherical cut-off distance R_{cut}, it is possible to determine the number of a given molecular or atomic species coordinated to a central ion (note that R_{cut} will be different for different molecules). Following the common practice we defined R_{cut} as the distance for which the corresponding radial distribution functions shows a first minimum. First, we have calculated the distribution of carbon coordination numbers around each ionic (BMI^+, PF_6^-) or solvent (acetonitrile) molecule adsorbed in nanoporous carbon electrodes held at various electrical potential. The values can differ markedly, ranging from 0 to 100, and the occurrence of several broad peaks indicates that the molecules have some preferred coordination numbers [107]. By examining typical configurations for each coordination number, we have defined four different adsorption modes for the molecules: They can be lying close to (1) an edge site, i.e. a carbon surface with a concave curvature, (2) a plane, which will have the local structure of a graphene sheet, (3) an hollow site, i.e. a carbon surface with a convex curvature and (4) a pocket, when they are inside a subnanometer carbon pore with a cylinder-like shape. Typical configurations corresponding to edge, plane and hollow sites are shown in the lower panel of Fig. 6.

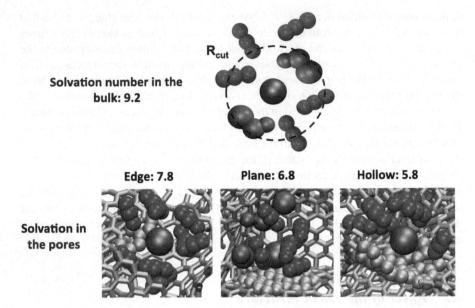

Fig. 6 *Top* Around each molecular ion, the number of neighbours of each species kind is determined using a spherical cut-off distance R_{cut}. In this example, the central anion (*green sphere*) is surrounded by two cations (*red spheres*) and three acetonitrile molecules (*blue spheres*). Two other solvent molecules have their centre of mass behind R_{cut}, they are therefore not considered to be in the solvation shell of the anion. *Bottom* Representative configurations for the edge, plane and hollow adsorption modes (turquoise rods: C–C bonds, *red* BMI$^+$, *green* PF$_6^-$, *blue* acetonitrile, turquoise: C atoms which are in the coordination sphere of the central anion). The average solvation number is also provided for each case (Color figure online)

In addition, as expected from experimental studies [110, 115], the adsorption of ions in confined environments leads to their partial desolvation. This is clearly visible on the average number of acetonitrile solvent molecules around each ion (solvation number), as shown in Fig. 6 for PF$_6^-$. In the bulk electrolyte, this average solvation number is of 9.2 (which corresponds to a distribution centred around 9, but it is worth noting that occurrences of solvation numbers down to 4 and up to 14 are also observed). The more confined, the less solvated they become, with average coordination numbers of 7.8, 6.8 and 5.8 for edge, planar and hollow sites, respectively. In the case of pocket sites, we were not able to extract an average solvation number due to the lack of statistics. Such a large desolvation effect has already been observed experimentally for aqueous solutions of RbBr by Ohkubo et al. [116] who have showed that Rb$^+$ and Br$^-$ hydration numbers decrease in slit-shape carbon nanospace by using extended X-ray absorption fine structure spectroscopy.

The different solvation numbers, associated with various pore geometries, are tightly linked to the efficiency of charge storage. A higher desolvation will generally lead to a higher charge on the carbon atoms coordinating an ion [107]. These local arrangements provide an explanation for the different capacitances which can be observed for different porous carbons with the same average pore size.

4.3 Dynamics of Charging: Coarse-Graining Further

So far most MD studies on supercapacitors have focused on the capacitance, without addressing the dynamic aspects. Nevertheless, it is the transport of the ions inside the pores which will control the power density of a device. It is thus of primary importance to characterize this transport on the molecular scale. A few studies have investigated the relaxation dynamics of ions close to planar electrodes [83, 118, 119]. In a study directed at the nanoporous carbons, Kondrat and Kornyshev have recently observed by combining a mean-field model with MD simulations that the charging of initially empty pores proceeds in a front-like way, while that of already filled pores is diffusive [120, 121]. In order to study the specific case of CDC nanoporous electrodes, we have performed a series of simulations in which, after equilibration of the system at 0 V, we suddenly apply an electric potential difference and follow the charging dynamics of the electrodes [117].

At the nanometer scale, strong heterogeneities are observed due to the particular structure of CDCs. These materials consist in a complex network of pores of various sizes, and differ substantially from the carbon nanotubes or the slit pores which are often used to model them. Depending on the local organization of the porous networks, some regions may start to charge before others despite being located more deeply inside the electrode. For example, in one of the carbon structures we have studied, a region in the middle of the electrode charges very quickly because it is located near a large pore of the carbon.

By comparing three CDCs with different structures we have shown that the charging time depends on the average pore size. CDC-1200 and CDC-950, which have similar average pore sizes, are charged on similar timescales, even though the details of the pore connectivity and the distribution of pore throats may also play a role. The process is slower by a factor of 4–8 in the case of CDC-800, which has a smaller average pore size. Experimental supercapacitors involving nanoporous carbon show good power performances in addition to their excellent energy densities [7], a result which seemed rather counterintuitive and had not been explained at the microscopic scale previously.

From our MD simulations data, which are shown in Fig. 7 for one of the carbon structures, we have fitted a macroscopic model based on an equivalent electric circuit. The transmission line model which is used in experiments is clearly able to capture the charging behaviour. This fit provides us with a value for the resistance of the electrolyte inside the pores. This quantity can then be used to calculate the charging times of experimental devices, which usually have a thickness of approximately 100 µm. We obtain characteristic charging times ranging between 2 and 8 s (depending on the carbon structure), which is of the correct order of

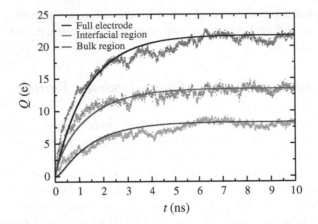

Fig. 7 Total charge of the full electrode as a function of time for a CDC-based supercapacitor using an ionic liquid electrolyte. The line corresponds to a fit obtained using an equivalent circuit model [117]. The latter is also able to capture well the evolution of the charge in two different regions of the electrode, the interface with the ionic liquid on the one hand and the bulkier part on the other hand

magnitude compared to experiments [7]. This confirms that the transport of the ions is not much affected in the porous materials. In another study, Kondrat et al. [121] have also shown the existence of collective transport effects, which could enhance by one to three orders of magnitudes the charging times. If the existence of such a mechanism leading to fast transport of the charge is shown to be universal for a series of electrolytes and electrode structures, it means that nanoporous carbon-based supercapacitors have inherently high power densities, so that it is important that studies about new carbon materials mainly focus on enhancing their energy density.

5 Perspectives

The use of classical MD simulations to study electrochemical systems remains very scarce. This is due to the difficulty of modelling redox reactions, and for this reason most of the studies concern systems in which they do not occur. Indeed, in the case of Li-ion batteries, most of the studies have focused on the diffusion of lithium in electrolytes. The methods we have developed for studying constant voltage electrodes may be extended further for the studies of batteries electrodes, although much more complex simulation methodologies will have to be designed.

In parallel, the rise of ab initio MD and the continuous increase of the available computational power will probably open the way to rigorous studies of the dynamics inside Li-ion batteries materials. People will face similar problems as those currently observed in static DFT calculations: Choice of the functionals, of the correct system sizes, inclusion of dispersion interactions, etc. Nevertheless, if they can be overcome efficiently, such simulations will be able to provide predictions of most of the needed quantities, such as diffusion coefficients, electrical conductivities, preferential diffusion pathways, operating voltages.

The situation differs markedly for carbon materials-based supercapacitors: In that case the charge storage occurs through a reversible adsorption of ions on the surface of an electrode, and no redox reactions occur. There are therefore many MD studies on these systems, although very few of them include a realistic representation of the charge rearrangements which occur inside the metallic electrodes. In the past 5 years, MD simulations have been able to provide a quantitative understanding on the increase of the capacitance in nanopores. More recently, the question of the dynamics of charging has also been addressed, so that it is now possible to examine the power and energy density of a given setup via molecular simulations.

The next challenge will consist in adopting predictive approaches, in order to propose the best electrode/electrolyte combination. This will be a very hard task, given the numerous carbon structures which are continuously being synthesized and the hundreds of ionic liquids which are now available. In addition, several additional aspects will have to be included in the simulations, such as allowing the carbon structure to breathe during the simulation (small volume changes are generally observed in experiments). In our case, the use of coarse-grained models will probably not be sufficient for making accurate predictions; for many systems all-atom force fields will have to be used (and the polarization effects will have to be included for an accurate prediction of the dynamics).

References

1. Manthiram A (2011) Materials challenges and opportunities of lithium ion batteries. J Phys Chem Lett 2:176–184
2. Dunn B, Kamath H, Tarascon J-M (2011) Electrical energy storage for the grid: a battery of choices. Science 334:928–935
3. Mueller T, Hautier G, Jain A, Ceder G (2011) Evaluation of tavorite-structured cathode materials for lithium-ion batteries using high-throughput computing. Chem Mater 23: 3854–3862
4. Delacourt C, Ati M, Tarascon J-M (2011) Measurement of lithium diffusion coefficient in Li_yFeSO_4F. J Electrochem Soc 158:A741–A749
5. Barpanda P, Chotard J-N, Delacourt C, Reynaud M, Filinchuk Y, Armand M, Deschamps M, Tarascon J-M (2011) $LiZnSO_4F$ made in an ionic liquid: a ceramic electrolyte composite for solid-state lithium batteries. Angew Chem Int Ed 50:2526–2531
6. Malik R, Burch D, Bazant M, Ceder G (2010) Particle size dependence of the ionic diffusivity. Nano Lett 10:4123–4127
7. Chmiola J, Yushin G, Gogotsi Y, Portet C, Simon P, Taberna PL (2006) Anomalous increase in carbon capacitance at pore sizes less than 1 nanometer. Science 313:1760–1763
8. Wang H, Forse AC, Griffin JM, Trease NM, Trognko L, Taberna P-L, Simon P, Grey C (2013) In situ NMR spectroscopy of supercapacitors: insight into the charge storage mechanism. J Am Chem Soc 135:18968–18980

9. Forse AC, Griffin JM, Wang H, Trease NM, Presser V, Gogotsi Y, Simon P, Grey C (2013) Nuclear magnetic resonance study of ion adsorption on microporous carbide-derived carbon. Phys Chem Chem Phys 15:7722–7730

10. Levi MD, Salitra G, Levy N, Aurbach D, Maier J (2009) Application of a quartz-crystal microbalance to measure ionic fluxes in microporous carbons for energy storage. Nat Mater 8:872–875

11. Levi MD, Sigalov S, Salitra G, Elazari R, Aurbach D (2011) Assessing the solvation numbers of electrolytic ions confined in carbon nanopores under dynamic charging conditions. J Phys Chem Lett 2:120–124

12. Tsai W-Y, Taberna P-L, Simon P (2014) Electrochemical quartz crystal microbalance (EQCM) study of ion dynamics in nanoporous carbons. J Am Chem Soc 136:8722–8728

13. Mo Y, Ong SP, Ceder G (2012) First principles study of the $Li_{10}GeP_2S_{12}$ lithium super ionic conductor material. Chem Mater 24:15–17

14. Ong SP, Mo Y, Richards WD, Miara L, Lee HS, Ceder G (2013) Phase stability, electrochemical stability and ionic conductivity of the $Li_{10\pm1}MP_2X_{12}$ (M = Ge, Si, Sn, Al or P, and X = O, S or Se) family of superionic conductors. Energy Environ Sci 6:148–156

15. Ramzan M, Lebègue S, Kang TW, Ahuja R (2011) Hybrid density functional calculations and molecular dynamics study of lithium fluorosulphate, a cathode material for lithium-ion batteries. J Phys Chem C 115:2600–2603

16. Marrocchelli D, Madden PA, Norberg ST, Hull S (2009) Cation composition effects on oxide conductivity in the $Zr_2Y_2O_7$-Y_3NbO_7 system. J Phys Condens Matter 21:405403

17. Marrocchelli D, Madden PA, Norberg ST, Hull S (2011) Structural disorder in doped zirconias, part II: vacancy ordering effects and the conductivity maximum. Chem Mater 23:1365–1373

18. Norberg ST, Hull S, Ahmed I, Eriksson SG, Marrocchelli D, Madden PA, Li P, Irvine JTS (2011) Structural disorder in doped zirconias, part I: the $Zr_{0.8}Sc_{0.2}Y_xO_{1.9}$ (0.0 < x < 0.2) system. Chem Mater 23:1356–1364

19. Burbano M, Norberg ST, Hull S, Eriksson SG, Marrocchelli D, Madden PA, Watson GW (2012) Oxygen vacancy ordering and the conductivity maximum in Y_2O_3-doped CeO_2. Chem Mater 24:222–229

20. Burbano M, Nadin S, Marrocchelli D, Salanne M, Watson GW (2014) Ceria co-doping: synergistic or average effect? Phys Chem Chem Phys 16:8320–8331

21. Madden PA, Heaton RJ, Aguado A, Jahn S (2006) From first-principles to material properties. J Mol Struct THEOCHEM 771:9–18

22. Becker CA, Tavazza F, Trautt ZT, Buarque de Macedo RA (2013) Considerations for choosing and using force fields and interatomic potentials in materials science and engineering. Curr Opin Solid State Mat Sci 17:277–283

23. Russo MF Jr, van Duin ACT (2011) Atomistic-scale simulations of chemical reactions: bridging from quantum chemistry to engineering. Nucl Instrum Methods Phys Res Sect B 269:1549–1554

24. Tang KT, Toennies JP (1984) An improved simple model for the van der Waals potential based on universal damping functions for the dispersion coefficients. J Chem Phys 80: 3726–3741

25. Aguado A, Madden PA (2003) Ewald summation of electrostatic multipole interactions up to the quadrupolar level. J Chem Phys 119:7471–7483

26. Reed SK, Lanning OJ, Madden PA (2007) Electrochemical interface between an ionic liquid and a model metallic electrode. J Chem Phys 126:084704

27. Siepmann JI, Sprik M (1995) Influence of surface-topology and electrostatic potential on water electrode systems. J Chem Phys 102:511–524

28. Merlet C, Péan C, Rotenberg B, Madden PA, Simon P, Salanne M (2013) Simulating supercapacitors: can we model electrodes as constant charge surfaces? J Phys Chem Lett 4:264–268

29. Kolafa J (2004) Time-reversible always stable predictor-corrector method for molecular dynamics of polarizable molecules. J Comput Chem 25:335–342

30. Kühne TD, Krack M, Mohamed FR, Parrinello M (2007) Efficient and accurate Car-Parrinello-like approach to Born-Oppenheimer molecular dynamics. Phys Rev Lett 98:066401
31. Roy D, Maroncelli M (2010) An improved four-site ionic liquid model. J Phys Chem B 114:12629–12631
32. Edwards DMF, Madden P, McDonald I (1984) A computer simulation study of the dielectric properties of a model of methyl cyanide. Mol Phys 51:1141–1161
33. Armand M, Tarascon J-M (2008) Building better batteries. Nature 451:652–657
34. Padhi AK, Nanjundaswamy KS, Goodenough JB (1997) Phospho-olivines as positive-electrode materials for rechargeable lithium batteries. J Electrochem Soc 144:1188–1194
35. Delmas C, Maccario M, Croguennec L, Le Cras F, Weill F (2008) Lithium deintercalation in LiFePO$_4$ nanoparticles via a domino-cascade model. Nat Mater 7:665–671
36. Sun C, Rajasekhara S, Goodenough JB, Zhou F (2011) Monodisperse porous LiFePO$_4$ microspheres for a high power li-ion battery cathode. J Am Chem Soc 133:2132–2135
37. Nishimura S-I, Nakamura M, Natsui R, Yamada A (2010) New lithium iron pyrophosphate as 3.5 V class cathode material for lithium ion battery. J Am Chem Soc 132:13596–13597
38. Clark JM, Nishimura S-I, Yamada A, Islam MS (2012) High-voltage pyrophosphate cathode: insights into local structure and lithium-diffusion pathways. Angew Chem Int Ed 51:13149–13153
39. Armstrong AR, Lyness C, Panchmatia PM, Islam MS, Bruce PG (2011) The lithium intercalation process in the low-voltage lithium battery anode Li$_{1+x}$V$_{1-x}$O$_2$. Nat Mater 10:223–229
40. Eames C, Armstrong AR, Bruce PG, Islam MS (2012) Insights into changes in voltage and structure of Li$_2$FeSiO$_4$ polymorphs for lithium-ion batteries. Chem Mater 24:2155–2161
41. Recham N, Chotard J-N, Dupont L, Delacourt C, Walker W, Armand M, Tarascon J-M (2010) A 3.6 V lithium-based fluorosulphate insertion positive electrode for lithium-ion batteries. Nat Mater 9:68–74
42. Barpanda P, Ati M, Melot BC, Rousse G, Chotard J-N, Doublet M-L, Sougrati MT, Corr SA, Jumas J-C, Tarascon J-M (2011) A 3.90 V iron-based fluorosulphate material for lithium-ion batteries crystallizing in the triplite structure. Nat Mater 10:772–779
43. Kamaya N, Homma K, Yamakawa Y, Hirayama M, Kanno R, Yonemura M, Kamiyama T, Kato Y, Hama S, Kawamoto K, Mitsui A (2011) A lithium superionic conductor. Nat Mater 10:682–686
44. Malik R, Zhou F, Ceder G (2011) Kinetics of non-equilibrium lithium incorporation in LiFePO$_4$. Nat Mater 10:587–590
45. Armstrong AR, Kuganathan N, Islam MS, Bruce PG (2011) Structure and lithium transport pathways in Li$_2$FeSiO$_4$ cathodes for lithium batteries. J Am Chem Soc 133:13031–13035
46. Tripathi R, Gardiner GR, Islam MS, Nazar LF (2011) Alkali-ion conduction paths in LiFeSO$_4$F and NaFeSO$_4$F tavorite-type cathode materials. Chem Mater 23:2278–2284
47. Castets A, Carlier D, Zhang Y, Boucher F, Marx N, Croguennec L, Menetrier M (2011) Multinuclear NMR and DFT calculations on the LiFePO$_4$·OH and FePO$_4$·H$_2$O homeotypic phases. J Phys Chem C 115:16234–16241
48. Castets A, Carlier D, Zhang Y, Boucher F, Menetrier M (2012) A DFT-based analysis of the NMR fermi contact shifts in tavorite-like LiMPO$_4$·OH and MPO$_4$·H$_2$O (M = Fe, Mn, V). J Phys Chem C 116:18002–18014
49. Ben Yahia M, Lemoigno F, Rousse G, Boucher F, Tarascon J-M, Doublet M-L (2012) Origin of the 3.6 V to 3.9 V voltage increase in the LiFeSO$_4$F cathodes for Li-ion batteries. Energy Environ Sci 5:9584–9594
50. Morgan BJ, Watson GW (2010) GGA+ U description of lithium intercalation into anatase TiO$_2$. Phys Rev B 82:144119
51. Morgan BJ, Madden PA (2012) Lithium intercalation into TiO$_2$(B): a comparison of LDA, GGA, and GGA+ U density functional calculations. Phys Rev B 86:035147

52. Islam MS, Driscoll DJ, Fisher CAJ, Slater PR (2005) Atomic-scale investigation of defects, dopants, and lithium transport in the $LiFePO_4$ olivine-type battery material. Chem Mater 17:5085–5092
53. Salanne M, Siqueira LJA, Seitsonen AP, Madden PA, Kirchner B (2012) From molten salts to room temperature ionic liquids: simulation studies on chloroaluminate systems. Faraday Discuss 154:171–188
54. Burbano M, Marrocchelli D, Yildiz B, Tuller HL, Norberg ST, Hull S, Madden PA, Watson GW (2011) A dipole polarizable potential for reduced and doped CeO_2 obtained from first principles. J Phys Condens Matter 23:255402
55. Salanne M, Marrocchelli D, Watson GW (2012) Cooperative mechanism for the diffusion of Li^+ ions in $LiMgSO_4F$. J Phys Chem C 116:18618–18625
56. Marrocchelli D, Salanne M, Watson GW (2013) Effects of Li-ion vacancies on the ionic conduction mechanism of $LiMgSO_4F$. Modell Simul Mater Sci Eng 21:074003
57. Muller RP, Schultz PA (2013) Modelling challenges for battery materials and electrical energy storage. Modell Simul Mater Sci Eng 21:070301
58. Heaton RJ, Brookes R, Madden PA, Salanne M, Simon C, Turq P (2006) A first-principles description of liquid BeF_2 and its mixtures with LiF: 1. Potential development and pure BeF_2. J Phys Chem B 110:11454–11460
59. Salanne M, Simon C, Turq P, Heaton RJ, Madden PA (2006) A first-principles description of liquid BeF_2 and its mixtures with LiF: 2. Network formation in LiF-BeF_2. J Phys Chem B 110:11461–11467
60. Marrocchelli D, Salanne M, Madden PA, Simon C, Turq P (2009) The construction of a reliable potential for GeO_2 from first principles. Mol Phys 107(4–6):443–452
61. Marrocchelli D, Salanne M, Madden PA (2010) High-pressure behaviour of GeO_2: a simulation study. J Phys Condens Matter 22:152102
62. Salanne M, Rotenberg B, Simon C, Jahn S, Vuilleumier R, Madden PA (2012) Including many-body effects in models for ionic liquids. Theor Chem Acc 131:1143
63. Rotenberg B, Salanne M, Simon C, Vuilleumier R (2010) From localized orbitals to material properties: building classical force fields for nonmetallic condensed matter systems. Phys Rev Lett 104:138301
64. Frenkel D, Smit B (2002) Understanding molecular dynamics, 2nd edn. Academic Press, Waltham
65. Chroneos A, Parfitt D, Kilner JA, Grimes RW (2010) Anisotropic oxygen diffusion in tetragonal $La_2NiO_{4+\delta}$: molecular dynamics calculations. J Mater Chem 20:266–270
66. Kushima A, Parfitt D, Chroneos A, Yildiz B, Kilner JA, Grimes RW (2011) Interstitialcy diffusion of oxygen in tetragonal $La_2CoO_{4+\delta}$. Phys Chem Chem Phys 13:2242–2249
67. Panchmatia PM, Orera A, Rees GJ, Smith ME, Hanna JV, Slater PR, Islam MS (2011) Oxygen defects and novel transport mechanisms in apatite ionic conductors: combined ^{17}O NMR and modeling studies. Angew Chem Int Ed 50:9328–9333
68. Murch GE (1982) The Haven ratio in fast ionic conductors. Solid State Ionics 7:177–198
69. Castiglione MJ, Madden PA (2001) Fluoride ion disorder and clustering in superionic PbF_2. J Phys Condens Matter 13:9963
70. Sebastian L, Gopalakrishnan J, Piffard Y (2002) Synthesis, crystal structure and lithium ion conductivity of $LiMgFSO_4$. J Mater Chem 12:374–377
71. Kilner JA (2000) Fast oxygen transport in acceptor doped oxides. Solid State Ionics, 129:13–23. (11th International conference on solid state ionics (SSI-11), Honolulu, Hawaii, 16–21 Nov 1997)
72. Raymundo-Piñero E, Kierzek K, Machnikowski J, Béguin F (2006) Relationship between the nanoporous texture of activated carbons and their capacitance properties in different electrolytes. Carbon 44:2498–2507
73. Largeot C, Portet C, Chmiola J, Taberna PL, Gogotsi Y, Simon P (2008) Relation between the ion size and pore size for an electric double-layer capacitor. J Am Chem Soc 130: 2730–2731
74. Simon P, Gogotsi Y (2008) Materials for electrochemical capacitors. Nat Mater 7:845–854

75. Kornyshev AA (2007) Double-layer in ionic liquids: paradigm change? J Phys Chem B 111:5545–5557
76. Bazant MZ, Storey BD, Kornyshev AA (2011) Double layer in ionic liquids: overscreening versus crowding. Phys Rev Lett 106:046102
77. Alam MT, Islam MM, Okajima T, Ohsaka T (2007) Measurements of differential capacitance in room temperature ionic liquid at mercury, glassy carbon and gold electrode interfaces. Electrochem Commun 9:2370–2374
78. Islam MM, Alam MT, Ohsaka T (2008) Electrical double-layer structure in ionic liquids: a corroboration of the theoretical model by experimental results. J Phys Chem C 112:16568–16574
79. Islam MM, Alam MT, Okajima T, Ohsaka T (2009) Electrical double layer structure in ionic liquids: an understanding of the unusual capacitance-potential curve at a nonmetallic electrode. J Phys Chem C 113:3386–3389
80. Alam MT, Islam MM, Okajima T, Ohsaka T (2009) Electrical double layer in mixtures of room-temperature ionic liquids. J Phys Chem C 113:6596–6601
81. Silva F, Gomes C, Figueiredo M, Costa R, Martins A, Pereira CM (2008) The electrical double layer at the [BMIM][PF$_6$] ionic liquid/electrode interface—effect of temperature on the differential capacitance. J Electroanal Chem 622:153–160
82. Heyes DM, Clarke JHR (1981) Computer-simulation of molten-salt interphases—effect of a rigid boundary and an applied electric-field. J Chem Soc Faraday Trans 2(77):1089–1100
83. Lanning O, Madden PA (2004) Screening at a charged surface by a molten salt. J Phys Chem B 108:11069–11072
84. Fedorov MV, Kornyshev AA (2008) Ionic liquid near a charged wall: structure and capacitance of electrical double layer. J Phys Chem B 112:11868–11872
85. Fedorov MV, Kornyshev AA (2008) Towards understanding the structure and capacitance of electrical double layer in ionic liquids. Electrochim Acta 53:6835–6840
86. Fedorov MV, Georgi N, Kornyshev AA (2010) Double layer in ionic liquids: the nature of the camel shape of capacitance. Electrochem Commun 12:296–299
87. Georgi N, Kornyshev AA, Fedorov MV (2010) The anatomy of the double layer and capacitance in ionic liquids with anisotropic ions: electrostriction vs. lattice saturation. J Electroanal Chem 649:261–267
88. Pounds M, Tazi S, Salanne M, Madden PA (2009) Ion adsorption at a metallic electrode: an *ab initio* based simulation study. J Phys Condens Matter 21:424109
89. Tazi S, Salanne M, Simon C, Turq P, Pounds M, Madden PA (2010) Potential-induced ordering transition of the adsorbed layer at the ionic liquid/electrified metal interface. J Phys Chem B 114:8453–8459
90. Vatamanu J, Borodin O, Smith GD (2010) Molecular dynamics simulations of atomically flat and nanoporous electrodes with a molten salt electrolyte. Phys Chem Chem Phys 12: 170–182
91. Kislenko SA, Samoylov IS, Amirov RH (2009) Molecular dynamics simulation of the electrochemical interface between a graphite surface and the ionic liquid [BMIM][PF$_6$]. Phys Chem Chem Phys 11:5584–5590
92. Feng G, Zhang JS, Qiao R (2009) Microstructure and capacitance of the electrical double layers at the interface of ionic liquids and planar electrodes. J Phys Chem C 113(11): 4549–4559
93. Merlet C, Salanne M, Rotenberg B, Madden PA (2011) Imidazolium ionic liquid interfaces with vapor and graphite: interfacial tension and capacitance from coarse-grained molecular simulations. J Phys Chem C 115:16613–16618
94. Merlet C, Salanne M, Rotenberg B (2012) New coarse-grained models of imidazolium ionic liquids for bulk and interfacial molecular simulations. J Phys Chem C 116:7687–7693
95. Feng G, Jiang D, Cummings PT (2012) Curvature effect on the capacitance of electric double layers at ionic liquid/onion-like carbon interfaces. J Chem Theory Comput 8:1058–1063

96. Yang L, Fishbine BH, Migliori A, Pratt LR (2009) Molecular simulation of electric double-layer capacitors based on carbon nanotube forests. J Am Chem Soc 131: 12373–12376

97. Shim Y, Kim HJ (2010) Nanoporous carbon supercapacitors in an ionic liquid: a computer simulation study. ACS Nano 4:2345–2355

98. Feng GA, Qiao R, Huang JS, Dai S, Sumpter BG, Meunier V (2011) The importance of ion size and electrode curvature on electrical double layers in ionic liquids. Phys Chem Chem Phys 13:1152–1161

99. Feng G, Li S, Atchison JS, Presser V, Cummings PT (2013) Molecular insights into carbon nanotube supercapacitors: capacitance independent of voltage and temperature. J Phys Chem C 117:9178–9186

100. David DM, Clarke JHR (1981) Computer simulation of molten-salt interphases. Effect of a rigid boundary and an applied electric field. J Chem Soc Faraday Trans 2(77):1089–1100

101. Atkin R, Warr GG (2007) Structure in confined room-temperature ionic liquids. J Phys Chem C 111:5162–5168

102. Hayes E, Warr GG, Atkin R (2010) At the interface: solvation and designing ionic liquids. Phys Chem Chem Phys 12:1709–1723

103. Atkin R, Borisenko N, Drüschler M, Zein El Abedin S, Endres F, Hayes R, Huber B, Roling B (2011) An in situ STM/AFM and impedance spectroscopy study of the extremely pure   1-butyl-1-methylpyrrolidinium   tris(pentafluoroethyl)trifluorophosphate/Au(111) interface: potential dependent solvation layers and the herringbone reconstruction. Phys Chem Chem Phys 13:6849–6857

104. Perkin S, Crowhurst L, Niedermeyer H, Welton T, Smith AM, Gosvami NN (2011) Self-assembly in the electrical double layer of ionic liquids. Chem Commun 47:6572–6574

105. Feng G, Huang J, Sumpter BG, Meunier V, Qiao R (2011) A "counter-charge layer in generalized solvents" framework for electrical double layers in neat and hybrid ionic liquid electrolytes. Phys Chem Chem Phys 13:14723–14734

106. Merlet C, Rotenberg B, Madden PA, Taberna P-L, Simon P, Gogotsi Y, Salanne M (2012) On the molecular origin of supercapacitance in nanoporous carbon electrodes. Nat Mater 11:306–310

107. Merlet C, Péan C, Rotenberg B, Madden PA, Daffos B, Taberna P-L, Simon P, Salanne M (2013) Highly confined ions store charge more efficiently in supercapacitors. Nat Commun 4:2701

108. Gogotsi Y, Nikitin A, Ye H, Zhou W, Fischer JE, Yi B, Foley HC, Barsoum MW (2003) Nanoporous carbide-derived carbon with tunable pore size. Nat Mater 2:591–594

109. Dash R, Chmiola J, Yushin G, Gogotsi Y, Laudisio G, Singer J, Fisher JE, Kucheyev S (2006) Titanium carbide derived nanoporous carbon for energy-related applications. Carbon 44:2489–2497

110. Chmiola J, Largeot C, Taberna P-L, Simon P, Gogotsi Y (2008) Desolvation of ions in subnanometer pores and its effect on capacitance and double-layer theory. Angew Chem Int Ed 47:3392–3395

111. Lin R, Huang P, Segalini J, Largeot C, Taberna PL, Chmiola J, Gogotsi Y, Simon P (2009) Solvent effect on the ion adsorption from ionic liquid electrolyte into sub-nanometer carbon pores. Electrochim Acta 54:7025–7032

112. Palmer JC, Llobet A, Yeon S-H, Fisher JE, Shi Y, Gogotsi Y, Gubbins KE (2010) Modeling the structural evolution of carbide-derived carbons using quenched molecular dynamics. Carbon 48:1116–1123

113. Willard AP, Chandler D (2010) Instantaneous liquid interfaces. J Phys Chem B 114: 1954–1958

114. Merlet C (2013) Modélisation de l'adsorption des ions dans les carbones nanoporeux

115. Ania CO, Pernak J, Stefaniak F, Raymundo-Piñero E, Béguin F (2009) Polarization-induced distortion of ions in the pores of carbon electrodes for electrochemical capacitors. Carbon 47:3158–3166

116. Ohkubo T, Konishi T, Hattori Y, Kanoh H, Fujikawa T, Kaneko K (2002) Restricted hydration structures of Rb and Br ions confined in slit-shaped carbon nanospace. J Am Chem Soc 124:11860–11861
117. Péan C, Merlet C, Rotenberg B, Madden PA, Taberna P-L, Daffos B, Salanne M, Simon P (2014) On the dynamics of charging in nanoporous carbon-based supercapacitors. ACS Nano 8:1576–1583
118. Pinilla C, Del Pópolo MG, Kohanoff J, Lynden-Bell RM (2007) Polarization relaxation in an ionic liquid confined between electrified walls. J Phys Chem B 111:4877–4884
119. Vatamanu J, Borodin O, Smith GD (2011) Molecular simulations of the electric double layer structure, differential capacitance, and charging kinetics for N-methyl-N-propylpyrrolidinium bis(fluorosulfonyl)imide at graphite electrodes. J Phys Chem B 115:3073–3084
120. Kondrat S, Kornyshev AA (2013) Charging dynamics and optimization of nano-porous supercapacitors. J Phys Chem C 117:12399–12406
121. Kondrat S, Wu P, Qiao R, Kornyshev AA (2014) Accelerating charging dynamics in subnanometre pores. Nat Mater 13:387–393

Continuum, Macroscopic Modeling of Polymer-Electrolyte Fuel Cells

Sivagaminathan Balasubramanian and Adam Z. Weber

Abstract In this chapter, the modeling equations and approaches for continuum modeling of phenomena in polymer-electrolyte fuel cells are introduced and discussed. Specific focus is made on the underlying transport, thermodynamic, and kinetic equations, and how these can be applied towards more complex fuel-cell issues such as multiphase flow. In addition, porous-media models including impact of droplets and pore-network modeling are introduced, as well methodologies towards modeling reaction rates in fuel-cell catalyst layers including physics-based impedance modeling. Finally, future directions for fuel-cell modeling are discussed.

1 Introduction

A polymer-electrolyte fuel-cell (PEFC) stack is a complex device involving different species in multiple phases distributed under varying temperature and pressure in all dimensions over a broad range of scales of dimensions from nanometers to hundreds of centimeters. Due to the coupling of multitude of processes, any change in one property in one cell may affect the reaction kinetics and consequently temperature distribution and fuel utilization within a stack as a whole for example. It is difficult to comprehend these effects and consequences on PEFC operation. In this aspect, mathematical modeling has played a significant role in PEFC technology development over the last 25 years. The models help in understanding different phenomena that enable technology development, in optimizing the system design for meeting the desired goals for practical applications such as power requirement or improved efficiency, and in controlling PEFC operation under dynamically varying load conditions.

S. Balasubramanian · A.Z. Weber (✉)
Environmental Energy Technologies Division, Lawrence Berkeley National Laboratory, Berkeley, CA, USA
e-mail: azweber@lbl.gov

© Springer-Verlag London 2016
A.A. Franco et al. (eds.), *Physical Multiscale Modeling and Numerical Simulation of Electrochemical Devices for Energy Conversion and Storage*, Green Energy and Technology, DOI 10.1007/978-1-4471-5677-2_4

The number of phases and components involving different physics and with widely varying operating conditions necessitates the use of mathematical modeling over different scales and dimensions. A wide range of modeling approaches—from nanoscale describing material or components to macroscale describing stack or system—have been developed. These models can be classified in many ways based on dimensionality, phenomena targeted for understanding, specific component, cell, or system level, etc. The continuum approach is mostly applied to cell-level models, which is traditionally where most of the focus of the modeling community lies. Component- and cell-level models describing major phenomena affecting PEFC operation have played and continue to play a major role in the underlying knowledge and design of PEFC operation and systems. Stack-level models play a significant role in controlling the dynamic PEFC operation in sync with other systems. While continuthe different phenomenaum modeling is the preferred choice for studying and understanding PEFC operation, recent focus also includes first-principles based approaches, such as molecular dynamics and ab-initio electronic structure, and mesoscale modeling of transport pathways and related phenomena at small length scales. While these models are discussed in detail in the other chapters of this book, this chapter focuses on the current progress in the macroscopic, continuum approach and how other modeling approaches can be employed collaboratively.

A robust model should predict PEFC behavior accurately under different operating conditions. Such robustness requires comprehensive description of all the major phenomena constituting PEFC operation. In practice, a comprehensive model describing PEFC operation under all the operating conditions is impossible because of the computational cost and limited understanding of the coupling of various phenomena. The general approach is to describe specific phenomena under limited conditions and gain better understanding of the system and then iteratively improve the coupling of the different phenomena. Computational efficiency balances the tradeoff between comprehensiveness and accuracy of model predictions. Therefore, the accuracy of model predictions is always limited by the choice of governing equations, input system parameters, and assumptions relaxing the complexity of the model. In this chapter, we will discuss about these factors and how to employ them effectively.

PEFC operation can be described by various interlinked processes, which, as shown in Fig. 1, are traditionally separated into kinetic, ohmic, and mass-transport losses. The major processes are mass and species transport coupled with reactions, momentum, charge, and thermal energy transport. Among them, some of the processes are usually limiting and determine the state of the PEFC at that instant. The state of the PEFC is predicted by solving the coupled system of equations governing these processes. For example, an increase in oxygen concentration increases the production of water, which, if not removed, will start blocking the oxygen transport and subsequently reduce the current and thus water generation. The objective of a model is to find the balance between the competing processes and consequently describe the state of the system, with an eye toward optimizing performance.

Fig. 1 Sample polarization curve showing dominant losses

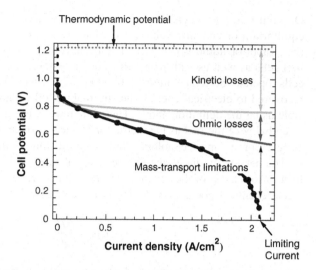

For metrics on performance and the state, the work performed by the system is simply the power as expressed by

$$P = IV \tag{1}$$

where I is the current and V is the cell voltage.

The potential corresponding to the Gibbs free energy is defined as the equilibrium or thermodynamic potential (see Fig. 1),

$$U^\theta = \frac{\Delta G}{z_i F} \tag{2}$$

Usually, most devices are operated not at the reference conditions. To account for this effect, one can determine from the first law of thermodynamics,

$$U = U^\theta + \Delta U = U^\theta + \frac{\Delta S}{z_i F}\left(T - T^\theta\right) \tag{3}$$

which can be calculated from handbooks [1, 2] for the electrochemical reaction of hydrogen with oxygen [18, 19]. Depending on if the product water is vapor or liquid, one arrives at different potentials due to the latent heat and free energy difference between liquid (U^θ) and vapor (U^*) water (the two potentials are related logarithmically by the vapor pressure of water, which is why they cross at 100 °C).

For PEFCs, the efficiency of the cell, η_{eff}, is typically defined relative to the maximum free energy available for electrical work,

$$\eta_{\text{eff}} = 1 - \frac{V}{U} \tag{4}$$

One must also be cognizant of whether the efficiency is defined in terms of the equilibrium or enthalpy values, and what the reference state is for the calculation (i.e., vapor or liquid water). This is especially important when comparing different fuel cells as well as with fuel cells to other systems. For example, solid-oxide fuel cells operate at temperatures (600–900 °C) where the heat generated can be recovered to electrical energy, thereby making efficiency greater than 100 % possible using the definition above. Thus, it is more advisable to use the heating-value or enthalpy of the fuel as the metric for efficiency since this also allows for a better comparison among technologies (e.g., combustion engines to fuel cells).

The net energy due to the electrochemical reaction is the difference between the heat of formation of the products and reactants, ΔH, which can be converted to an electrochemical potential, resulting in the enthalpy potential,

$$U_H = \frac{\Delta H}{z_i F} \tag{5}$$

where z_i is the charge number of species i and F is Faraday's constant. Thus, the expression for the heat released becomes

$$Q = i(U_H - V) \tag{6}$$

If the cell potential equals the enthalpy potential, there is no net heat loss, which is why the enthalpy potential is often termed the thermoneutral potential. However, the enthalpy energy is not fully accessible as it is composed of both reversible or entropic as well as irreversible components.

Though there are many works that attempt to find analytical solutions for the system of equations [3, 4], finding an analytical solution for the system of equations for all of the possible conditions is not viable. Hence, most of the models are numerically solved. The general approach is to segment the entire domain into multiple nodes, volumes, or elements and then solve the conservation laws numerically in each control segment. For improved numerical stability, the vector components are solved at the interface of the segments, and scalar quantities are balanced between neighboring segments (i.e., control-volume approach) [5].

The outline of this chapter is as follows. First, the general governing transport and conservation equations as applied to PEFCs are reviewed. Next, equations and the approach for the macroscopic modeling of the gas-diffusion media (GDM), membrane, and catalyst layers are presented. After a section on modeling electrochemical impedance spectroscopy (EIS), a summary including remaining challenges is made. Throughout the sections, current issues and needs are mentioned including those related to multiscaling. Before discussing specific equations, model dimensionality should be mentioned.

1.1 Modeling Dimension

Zero-dimensional (0-D) models relate system variables such as cell voltage, current, and temperature using simple empirical correlations without consideration of spatial domain [6–13]. A typical 0-D model equation for polarization curve is

$$V = U - b\log\left(\frac{i}{i_0}\right) - R'i + b\log\left(1 - \frac{i}{i_{\text{lim}}}\right) \tag{7}$$

and accounts for the major losses as shown in Fig. 1. The first term on the right corresponds to the thermodynamic cell potential. The second term represents the loss in cell potential to kinetic resistance where b is the Tafel slope and i_0 is the exchange current density. The third term accounts for the ohmic losses where R' is the total ohmic resistance (contact and cell). The last term represents the limiting current caused by concentration overpotential. As 0-D models do not provide fundamental understanding of PEFC operation, they are not really suitable for predicting performance for different operating conditions or optimizing the design. System-level models that see a PEFC as a black box use 0-D models for controlling the stack and system peripherals; they are used to correlate different stack properties with operating conditions.

   In stack-level models, component-level phenomena are ignored. These models focused on temperature and pressure change in the stack with load variation and how that may affect the coupling with other system-level components like compressors, reformers, etc. [14]. Stack-level management of PEFCs is the desired controlling method of PEFC operation as monitoring of intrinsic variables (if possible) might overload the control system and make it more complex. Stack-level models help in two ways: one is in understanding the balance of power and the other is to develop dynamic control systems. In stacks, the multiple cells and interactions typically necessitate the use of 0-D models. Thus, the performance is given by Eq. 7. For thermal management, a key issue for stack design, one can use

$$\Delta \dot{Q}_{\text{stack}} = \dot{Q}_{\text{in}} - \dot{Q}_{\text{out}} + \dot{Q}_{\text{gen}} = m_s c_{p,s}(T_s)\frac{dT_s}{dt} \tag{8}$$

where the heat generation within each cell can be given by Eq. 6 and the heat removal can be due to coolant flows or just natural convection. The thermal balance allows one to assess the balance of power and energy for startup, cooling channel requirements, and switching the reactants flow depending on the changing power-utilization conditions. The switching also helps in economic usage of the system to get maximum efficiency of the stack with minimal loss. It would also be appropriate to consider the cost of operational life while arriving at the operational condition of the stack using such simple models, especially if the cells and stacks have been well defined (i.e., the terms in Eq. 7 are measured for the exact system being studied).

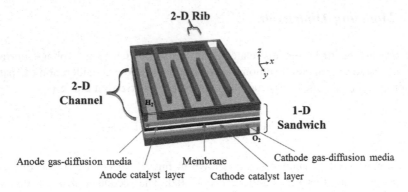

Fig. 2 Schematic of spatial domains and associated modeling including the PEFC sandwich

As shown in Fig. 2, one dimensional (1-D) models describe the physical phenomena typically in the through-plane direction [15–17]. Comprehensive 1-D models incorporate electrochemical reaction at the porous electrodes, transport of gas and liquid species through porous gas-diffusion media (GDM), and transport of charged species like electrons and protons. These models treat the cell as multiple layers bonded together. Proper interfacial internal boundary conditions are used to couple the different processes. Along-the-channel 1-D models focus on the transport and depletion of fuel and oxygen along the channel.

Two-dimensional models use the 1-D model direction and the other direction can be classified into across-the-channel or along-the-channel models [18]. Across-the-channel models focus on a cross-section of the flow channel including the rib and channel. This approach addresses the effects of the solid rib and channel. Along-the-channel 2-D models incorporate the effects of fuel and oxygen depletion and water accumulation on the current distribution along the channel and in the through-plane direction. This helps in understanding the different types of channel configuration and flow direction, which cannot be addressed with solely 1-D modeling. One can use arguments of spatial separation to assume that the conditions in the cell sandwich only propagate and interact along the channel and not internally. This finding led to a group of models referred to as pseudo-2-D models. Instead of solving the coupled conservation equations in a 2-D domain, the 1-D model is solved at each node along the channel, thereby reducing the computational cost of a full 2-D model.

3-D models are comprehensive models that describe all the axis of a PEFC. In addition to understanding distribution of species and current, these models show the distribution of temperature and fluid in a 3-D spatial domain, especially the effects of cooling channels, channel cross-section, and channel design. These models typically use computational fluid dynamics (CFD). Similar to pseudo 2-D models, there are also a class of models termed pseudo 3-D or 2-D + 1. In this formulation, the along-the-channel direction is only interacting at the boundaries between cell components, but instead of 1-D sandwich models, 2-D across-the-channel models are used.

Finally, multiscale models can be considered as being multidimensional. For example, there are 3-D models at the microscale that can determine properties to the macroscale model that may be a 1-D model. Also, porous electrodes (as discussed in Sect. 5.2) typically examine reaction into the reactive particle as well as across the domain, which can be considered two separate dimensions.

2 Basic Governing Equations

To model PEFC behavior at the continuum scale requires the use of overall conservation equations for mass, momentum, energy, and charge transport within the various subdomains. These equations are more or less known and used in their general forms, with the complications and multiscale aspects arising from the need to determine the correct boundary conditions and effective properties. In this section, the general conservation laws are presented along with the general, well-known transport equations. Later in this chapter, specific relations that link physical properties within the various subdomain classes are discussed along with modified forms of the general equations.

For a control segment, the general conservation equation for property ψ, representing any of the aforementioned transport processes, can be written as,

$$\frac{\partial \psi}{\partial t} + \nabla \cdot N_\psi = S_\psi \tag{9}$$

The first term represents the time-dependent property and is neglected for description of steady-state operation. While a vast majority of PEFC models are steady-state models, there are transient models that address specific transient phenomena such as, degradation mechanisms [19], contamination effects [20], load-change effects [21], start-stop cycles, and cold-start [22–24].

The second term in Eq. 9 represents the change in the property ψ due to flux (N) into or out of the control segment under study. The flux denotes the transfer of property driven by imbalances within the system and is the result of system adjusting itself to bring certain equilibrium. The spatial derivative of flux addresses the distribution of the property over the spatial domain.

The third term (S) called the source term represents all the processes that cause generation or decay of the property driven by an imbalance within the control segment. For example, a supersaturated vapor phase may condense within the control segment and leads to a decrease in vapor-phase concentration and an increase in liquid-phase concentration. This term incorporates all other terms such as reaction terms and phase-change terms that are not captured by the flux. The term couples different conservation laws within the segment.

2.1 Material

The conservation of material can be written as in Eq. 9 except that the physical quantity ψ could be p—partial pressure of gas, c—concentration of solution, x—mole fraction of particular species, or ρ—density of fluid. However, for the case of a mixture in a multiphase system, it is necessary to write material balances for each of the component in each phase k, which in summation can still govern the overall conservation of material,

$$\frac{\partial \varepsilon_k c_{i,k}}{\partial t} = -\nabla \cdot \mathbf{N}_{i,k} - \sum_h a_{1,k} s_{i,k,h} \frac{i_{h,1-k}}{n_h F} + \sum_l s_{i,k,l} \sum_{p \neq k} a_{k,p} r_{l,k-p} + \sum_g s_{i,k,g} \varepsilon_k R_{g,k}$$

(10)

In the above expression, ε_k is the volume fraction of phase k, $c_{i,k}$ is the concentration of species i in phase k, and $s_{i,k,l}$ is the stoichiometric coefficient of species i in phase k participating in heterogeneous reaction l, $a_{k,p}$ is the specific surface area (surface area per unit total volume) of the interface between phases k and p, $i_{h,l-k}$ is the normal interfacial current transferred per unit interfacial area across the interface between the electronically conducting phase and phase k due to electron-transfer reaction h, and is positive in the anodic direction.

The term on the left side of the equation is the accumulation term, which accounts for the change in the total amount of species i held in phase k within a differential control volume over time. The first term on the right side of the equation keeps track of the material that enters or leaves the control volume by mass transport as discussed in later sections. The remaining three terms account for material that is gained or lost (i.e., source terms, S_ψ, in Eq. 9). The first summation includes all electron-transfer reactions that occur at the interface between phase k and the electronically conducting phase l; the second summation accounts for all other interfacial processes that do not include electron transfer like evaporation or condensation; and the final term accounts for homogeneous chemical reactions in phase k. It should be noted that in terms of an equation count, for n species there are only $n - 1$ conservation equations needed since one can be replaced by

$$\sum_i x_i = 1$$

(11)

In the above material balance (Eq. 10), one needs an expression for the flux or transport of material. Often, this expression stems from considering only the interactions of the various species with the solvent

$$\mathbf{N}_{i,k} = -D_i \nabla c_{i,k} + c_i \mathbf{v}_k$$

(12)

where $\mathbf{v}_k$ is the mass-averaged velocity of phase k

$$\mathbf{v}_k = \frac{\sum_{i \neq s} M_i \mathbf{N}_{i,k}}{\rho_k} \tag{13}$$

One can see that if convection is neglected, Eq. 12 results in Fick's law. Substitution of Eq. 12 into Eq. 10 results in the equation for convective diffusion,

$$\frac{\partial(\varepsilon_k \rho_k w_i)}{\partial t} + \nabla \cdot (\mathbf{v}_k \rho_k w_i) = \nabla \cdot \left(\rho_k D_i^{\text{eff}} \nabla w_i \right) + S_i \tag{14}$$

which is often used in CFD simulations. In the above expression, the reaction source terms are not shown explicitly, w_i is the mass fraction of species i, and the superscript 'eff' is used to denote an effective diffusion coefficient due to different phenomena or phases as discussed later in this chapter.

If the interactions among the various species are important, then Eq. 12 needs to be replaced with the multicomponent Stefan–Maxwell equations that account for binary interactions among the various species

$$\nabla x_{i,k} = -\frac{x_{i,k}}{RT} \left(\bar{V}_i - \frac{M_i}{\rho_k} \right) \nabla p_k + \sum_{j \neq i} \frac{x_{i,k} \mathbf{N}_{j,k} - x_{j,k} \mathbf{N}_{i,k}}{\varepsilon_k c_{T,k} D_{i,j}^{\text{eff}}} \tag{15}$$

where x_i and M_i are the mole fraction and molar mass of species i, respectively, $D_{i,j}^{\text{eff}}$ is the effective binary diffusion coefficient between species i and j, and $c_{T,k}$ is the total concentration of species in phase k as derived from the ideal-gas law. The first term on the right-side accounts for pressure diffusion (e.g., in centrifugation) which often can be ignored, but, on the anode side, the differences between the molar masses of hydrogen and water means that it can become important in certain circumstances [25]. The second term on the right side stems from the binary collisions between various components. For a multicomponent system, Eq. 15 results in the correct number of transport properties that must be specified to characterize the system, $1/2n(n-1)$, where n is the number of components and the $1/2$ is because $D_{i,j}^{\text{eff}} = D_{j,i}^{\text{eff}}$ by the Onsager reciprocal relationships.

The form of Eq. 15 is essentially an inverted form of the type of Eq. 12, since one is not writing the flux in terms of a material gradient but the material gradient in terms of the flux. This is not a problem, if one is solving the equations as written; however, many numerical packages require a second-order differential equation (e.g., see Eq. 14). To do this with the Stefan–Maxwell equations, inversion of them is required. For a two-component system where the pressure diffusion is negligible, one gets Eq. 12. For higher numbers of components, the inversion becomes cumbersome and analytic expressions are harder to obtain, resulting oftentimes in numerical inversion. In addition, the inversion results in diffusion coefficients which are more composition dependent. For example, Bird et al. show the form for a three-component system [26].

2.1.1 Charge

The conservation equation for charged species is an extension of the conservation of mass. Taking Eq. 10 and multiplying by $z_i F$ and summing over all species and phases while noting that all reactions are charge balanced yields

$$\frac{\partial}{\partial t} F \sum_k \sum_i z_i c_{i,k} = -\nabla \cdot F \sum_k \sum_i z_i \mathbf{N}_{i,k} \tag{16}$$

where the charge and current densities can be defined by

$$\rho_e = F \sum_k \sum_i z_i c_{i,k} \tag{17}$$

and

$$\mathbf{i}_k = F \sum_i z_i \mathbf{N}_{i,k}, \tag{18}$$

respectively. Because a large electrical force is required to separate charge over an appreciable distance, a volume element in the electrode will, to a good approximation, be electrically neutral; thus one can assume electroneutrality for each phase

$$\sum_i z_i c_{i,k} = 0 \tag{19}$$

The assumption of electroneutrality implies that the diffuse double layer, where there is significant charge separation, is small compared to the volume of the domain, which is normally (but not necessarily always) the case. The general charge balance (Eq. 16), assuming electroneutrality and the current definition (Eq. 18) becomes

$$\sum_k \nabla \cdot \mathbf{i}_k = 0 \tag{20}$$

While this relationship applies for almost all of the modeling, there are cases where electroneutrality does not strictly hold, including for some transients and impedance measurements, where there is charging and discharging of the double layer, as well as simulations at length scales within the double layer (typically on the order of nanometers) such as reaction models near the electrode surface. For these cases, the correct governing charge conservation results in Poisson's equation,

$$\nabla^2 \Phi = \frac{\rho_e}{\varepsilon_0} \tag{21}$$

where ε_0 is the permittivity of the medium. For the diffuse part of the double layer, often a Boltzmann distribution is used for the concentration of species i

$$c_i = c_{i,\infty} \exp\left(-\frac{z_i F \Phi}{RT}\right) \tag{22}$$

To charge this double layer, one can derive various expressions for the double-layer capacitance depending on the adsorption type, ionic charges, etc. [27], where the double-layer capacitance is defined as

$$C_d = \left(\frac{\partial q}{\partial \Phi}\right)_{\mu_i, T} \tag{23}$$

where q is the charge in the double layer and the differential is at constant composition and temperature. To charge the double layer, one can write an equation of the form

$$i = C_d \frac{\partial \Phi}{\partial t} \tag{24}$$

where the charging current will decay with time as the double layer becomes charged.

For the associated transport of charge, one can use either a dilute-solution or concentrated-solution approach. In general, the concentrated-solution approach is more rigorous but requires more knowledge of all of the various interactions (similar to the material-transport-equation discussion above). For the dilute-solution approach, one can use the Nernst–Planck equation,

$$\mathbf{N}_{i,k} = -z_i u_i F c_{i,k} \nabla \Phi_k - D_i \nabla c_{i,k} + c_{i,k} \mathbf{v}_k \tag{25}$$

where u_i is the mobility of species i. In the equation, the terms on the right side correspond to migration, diffusion, and convection, respectively. Multiplying Eq. 25 by $z_i F$ and summing over the species i in phase k,

$$F \sum_i z_i \mathbf{N}_{i,k} = -F^2 \sum_i z_i^2 u_i c_{i,k} \nabla \Phi_k - F \sum_i z_i D_i \nabla c_{i,k} + F \sum_i z_i c_{i,k} \mathbf{v}_k \tag{26}$$

and noting that the last term is zero due to electroneutrality (convection of a neutral solution cannot move charge) and using the definition of current density (Eq. 18), one gets

$$\mathbf{i}_k = -\kappa_k \nabla \Phi_k - F \sum_i z_i D_i \nabla c_{i,k} \tag{27}$$

where κ_k is the conductivity of the solution of phase k

$$\kappa_k = F^2 \sum_i z_i^2 c_{i,k} u_i \tag{28}$$

When there are no concentration variations in the solution, Eq. 27 reduces to Ohm's law,

$$\mathbf{i}_k = -\kappa_k \nabla \Phi_k \tag{29}$$

This dilute-solution approach does not account for interaction between the solute molecules. Also, this approach will either use too many or too few transport coefficients depending on if the Nernst–Einstein relationship is used to relate mobility and diffusivity,

$$D_i = RT u_i \tag{30}$$

which only rigorously applies at infinite dilution. Thus, the concentrated-solution theory approach is recommended, however, the Nernst–Planck equation can be used in cases where most of the transport properties are unknown or where the complex interactions and phenomena being investigated necessitate simpler equations (for example, transport of protons and water along a single-charged pore in the membrane).

The concentrated-solution approach for charge utilizes the same underpinnings as that of the Stefan–Maxwell equation, which starts with the original equation of multicomponent transport [28]

$$\mathbf{d}_i = c_i \nabla \mu_i = \sum_{j \neq i} K_{i,j} (\mathbf{v}_j - \mathbf{v}_i) \tag{31}$$

where $\mathbf{d}_i$ is the driving force per unit volume acting on species i and can be replaced by a electrochemical-potential gradient of species i, and $K_{i,j}$ are the frictional interaction parameters between species i and j. The above equation can be analyzed in terms of finding expressions for $K_{i,j}$'s, introducing the concentration scale including reference velocities and potential definition, or by inverting the equations and correlating the inverse friction factors to experimentally determined properties. Which route to take depends on the phenomena being studied (e.g., membrane, binary salt solution, etc.) [27] and will be examined in more detail below. If one uses a diffusion coefficient to replace the drag coefficients,

$$K_{i,j} = \frac{RT c_i c_j}{c_T \mathrm{D}_{i,j}} \tag{32}$$

where $\mathrm{D}_{i,j}$ is an interaction parameter between species i and j based on a thermo-dynamic driving force, then the multicomponent equations look very similar to the Stefan–Maxwell ones (Eq. 15). In addition, using the above definition for $K_{i,j}$ and assuming that species i is a minor component and that the total concentration, c_T, can be replaced by the solvent concentration (species 0), then Eq. 31 for species i in phase k becomes

$$\mathbf{N}_{i,k} = -\frac{D_{i,0}}{RT} c_{i,k} \nabla \mu_{i,k} + c_{i,k} \mathbf{v}_0 \tag{33}$$

This equation is very similar to the Nernst–Planck Equation (25), except that the driving force is the thermodynamic electrochemical potential, which contains both the migration and diffusive terms.

2.1.2 Momentum

Due to the highly coupled nature of momentum conservation and transport, both are discussed below. Also, the momentum or volume conservation equation is highly coupled to the mass or continuity conservation equation (Eq. 10). Newton's second law governs the conservation of momentum and can be written in terms of the Navier–Stokes equation [29]

$$\frac{\partial(\rho_k \mathbf{v}_k)}{\partial t} + \mathbf{v}_k \cdot \nabla(\rho_k \mathbf{v}_k) = -\nabla p_k + \mu_k \nabla^2 \mathbf{v}_k + S_m \tag{34}$$

where μ_k and $\mathbf{v}_k$ are the viscosity and mass-averaged velocity of phase k, respectively,

$$\mathbf{v}_k = \frac{\sum_{i \neq s} M_i \mathbf{N}_{i,k}}{\rho_k} \tag{35}$$

The transient term in the momentum conservation equation represents the accumulation of momentum with time and the second term describes convection of the momentum flux (which is often small for PEFC designs). The first two terms on the right side represent the divergence of the stress tensor and the last term represents other sources of momentum, typically other body forces like gravity or magnetic forces. For PEFCs, these forces are often ignored and unimportant, i.e. $S_m = 0$.

It should be noted that for porous materials, as discussed below, the Navier–Stokes equations are not used and instead one uses the more empirical Darcy's law for the transport equation [30, 31],

$$\mathbf{v}_k = -\frac{k_k}{\mu_k} \nabla p_k \tag{36}$$

This transport equation can be used as a first-order equation or combined with a material balance (Eq. 10) to yield

$$\frac{\partial(\rho_k \varepsilon_k)}{\partial t} = \nabla \cdot \left(-\rho_k \frac{k_k}{\mu_k} \nabla p_k\right) + S_m \tag{37}$$

which is similar to including it as a dominant source term. In the above expression, k_k is effective permeability of phase k.

2.1.3 Energy

Electrochemical reactions in a PEFC release electrical energy and heat energy as discussed in Sect. 1. While electrical energy is accounted by the flow of charge through the external circuit, the heat energy is conducted by the components and rejected either into cooling plates/channels or to the external environment. The heat causes an increase in the local temperature, which affects local properties and may also result in different transport mechanisms (e.g., phase-change-induced flow of water). Throughout all layers of the PEFC, the same transport and conservation equations exist for energy with the same general transport properties, and only the source terms vary.

The conservation of thermal energy can be written as

$$
\begin{aligned}
\rho_k \hat{C}_{p_k} &\left(\frac{\partial T_k}{\partial t} + \mathbf{v}_k \cdot \nabla T_k\right) + \left(\frac{\partial \ln \rho_k}{\partial \ln T_k}\right)_{p_k, x_{i,k}} \left(\frac{\partial p_k}{\partial t} + \mathbf{v}_k \cdot \nabla p_k\right) \\
&= Q_{k,p} - \nabla \cdot \mathbf{q}_k - \boldsymbol{\tau} : \nabla \mathbf{v}_k + \sum_i \bar{H}_{i,k} \nabla \cdot \mathbf{J}_{i,k} - \sum_i \bar{H}_{i,k} \Re_{i,k}
\end{aligned}
\tag{38}
$$

In the above expression, the first term represents the accumulation and convective transport of enthalpy, where $\hat{C}_{p_k}$ is the heat capacity of phase k which is a combination of the various components of that phase. The second term is energy due to reversible work. For condensed phases this term is negligible, and an order-of-magnitude analysis for ideal gases with the expected pressure drop in a PEFC demonstrates that this term is negligible compared to the others. The first two terms on the right side of Eq. 38 represent the net heat input by conduction and interphase transfer. The first is due to heat transfer between two phases

$$Q_{k,p} = h_{k,p} a_{k,p} (T_p - T_k) \tag{39}$$

where $h_{k,p}$ is the heat-transfer coefficient between phases k and p per interfacial area. Most often this term is used as a boundary condition since it occurs only at the edges. However, in some modeling domains (e.g., along the channel in Sect. 1.1) it may need to be incorporated as above. The second term is due to the heat flux in phase k

$$\mathbf{q}_k = -\sum_i \bar{H}_{i,k} \mathbf{J}_{i,k} - k_{T_k}^{\text{eff}} \nabla T_k \tag{40}$$

where $\bar{H}_{i,k}$ is the partial molar enthalpy of species i in phase k, $\mathbf{J}_{i,k}$ is the flux density of species i relative to the mass average velocity of phase k

$$\mathbf{J}_{i,k} = \mathbf{N}_{i,k} - c_{i,k} \mathbf{v}_k \tag{41}$$

and $k_{T_k}^{\text{eff}}$ is the effective thermal conductivity of phase k. The third term on the right side of Eq. 38 represents viscous dissipation, the heat generated by viscous forces, where $\boldsymbol{\tau}$ is the stress tensor. This term is also small, and for most cases can be neglected. The fourth term on the right side comes from enthalpy changes due to diffusion. Finally, the last term represents the change in enthalpy due to reaction

$$\sum_i \bar{H}_{i,k} \Re_{i,k} = -\sum_h a_{1,k} i_{h,1-k} \left(\eta_{s_{h,1-k}} + \Pi_h \right) - \sum_{p \neq k} \Delta H_l a_{k,p} r_{l,k-p} - \sum_g \Delta H_g R_{g,k} \tag{42}$$

where the expressions can be compared to those in the material conservation Eq. (10). The above reaction terms include homogeneous reactions, interfacial reactions (e.g., evaporation), and interfacial electron-transfer reactions. For the latter, the irreversible heat generation is represented by the activation overpotential and the reversible heat generation is represented by the Peltier coefficient, Π_h. The Peltier coefficient for charge-transfer reaction h and can be expressed as

$$\Pi_h \approx \frac{T}{n_h F} \sum_i s_{i,k,h} \bar{S}_{i,k} = T \frac{\Delta S_h}{n_h F} \tag{43}$$

where ΔS_h is the entropy of reaction h. The above equation neglects transported entropy (hence the approximate sign), and the summation includes all species that participate in the reaction (e.g., electrons, protons, oxygen, hydrogen, water). While the entropy of the half-reactions that occur at the catalyst layers is not truly obtainable since it involves knowledge of the activity of an uncoupled proton, the Peltier coefficients have been measured experimentally for these reactions, with most of the reversible heat due to the 4-electron oxygen reduction reaction [32, 33].

It is often the case that because of the intimate contact between the gas, liquid, and solid phases within the small pores of the various PEFC layers, that equilibrium can be assumed such that all of the phases have the same temperature as each other for a given point in the PEFC. Doing this eliminates the phase dependences in the above equations and allows for a single thermal energy equation to be written. Neglecting those phenomena that are minor as mentioned above and summing over the phases, results in

$$\sum_k \rho_k \hat{C}_{p_k} \frac{\partial T}{\partial t} = - \sum_k \rho_k \hat{C}_{p_k} \mathbf{v}_k \cdot \nabla T + \nabla \cdot \left(k_T^{\text{eff}} \nabla T \right) + \sum_k \frac{\mathbf{i}_k \cdot \mathbf{i}_k}{\kappa_k^{\text{eff}}}$$

$$+ \sum_h i_h (\eta_h + \Pi_h) - \sum_h \Delta H_h r_h \tag{44}$$

where the expression for Joule or ohmic heating has been substituted in from the third term in the right side of Eq. 38

$$- \sum_i \mathbf{J}_{i,k} \cdot \nabla \bar{H}_{i,k} = -\mathbf{i}_k \cdot \nabla \Phi_k = \frac{\mathbf{i}_k \cdot \mathbf{i}_k}{\kappa_k^{\text{eff}}} \tag{45}$$

In Eq. 44, the first term on the right side is energy transport due to convection, the second is energy transport due to condition, the third is the ohmic heating, the fourth is the reaction heats, and the last represents reactions in the bulk which include such things as vaporization/condensation and freezing/melting. Heat lost to the surroundings is only accounted for at the boundaries of the cell. In terms of magnitude, the major heat generation sources are the oxygen reduction reaction and the water phase changes, and the main mode of heat transport is through conduction.

3   Membrane

Polymer electrolytes are considered to be the heart of the PEFC. They facilitate the transport of ions between the electrodes while preventing the direct combustion of hydrogen with oxygen. The most studied ion-conducting membrane for PEFCs is a proton-conducting polymer composed of a chemically inert poly-tetrafluoroethylene (PTFE) backbone and side chains ending with hydrophilic sulfonic acid (SO_3^-) ionic groups, with the canonical one being Nafion® as shown in Fig. 3.

The main limitation of the state-of-the-art membrane is the need for high relative humidity to enable facile transport of protons; dry Nafion membrane does not conduct protons efficiently. The proton conductivity is not a fixed parameter of a membrane but a result of combination of several factors such as equivalent weight or ion-exchange capacity (a ratio of weight (g) of dry polymer to number of moles of acid group), temperature, mechanical stress, pretreatment, and contamination.

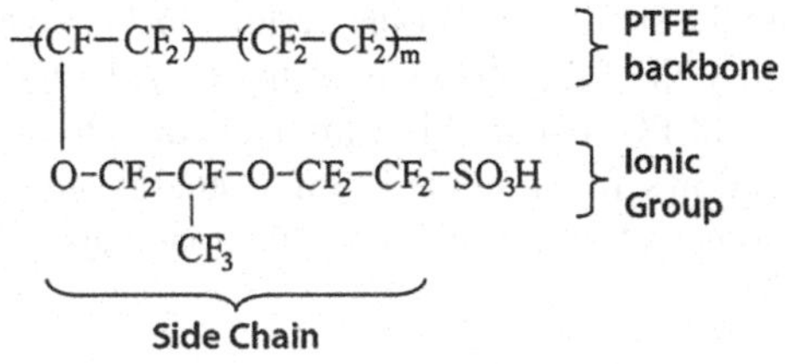

Fig. 3 Chemical structure of Nafion®

Fig. 4 Multiscale representation of phenomena of proton-exchange membranes (schematic courtesy of Ahmet Kusoglu)

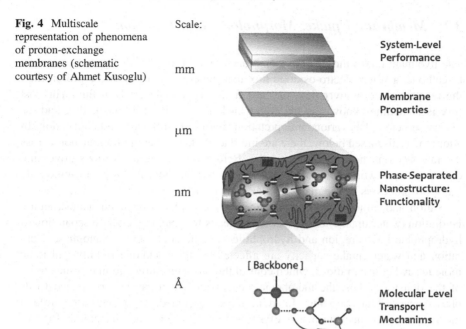

Considering these factors, a simple Ohm's law description, as done with solid electrical conductors, is not sufficient to treat this solid electrolyte medium; but a rigorous model incorporating these factors as variables is a necessity. Earlier models correlated the relative humidity with membrane water uptake and subsequently the proton conductivity of the membrane using mainly empirical expressions [16]. With the increase in the durability of PEFC, even minor structural effects are garnering attention. In addition, the advent of better diagnostics combined with the development of newer membranes with different ionomer moieties and morphology require better understanding of ion transport. This increased understanding and information has promoted a significant increase in ab initio, molecular dynamics, and coarse-grained approaches to predict transport and morphology. In principle, such models can inform the higher order continuum models either in terms of morphologies or effective transport properties; however, such linkages as shown in Fig. 4 are in their infancy and are an active area of current research. This section focuses on the continuum modeling to understand the key transport phenomena and associated water uptake and how it is employed in understanding overall PEFC operation.

3.1 Membrane Uptake, Morphology, and Function

The five main fluxes through the membrane are a proton flux that goes from anode to cathode, a water electro-osmotic flux that develops along with the proton flux, the reactant-gas crossover flux, the heat flux, and a water-gradient flux. This last flux is sometimes known as the water back flux or back-diffusion flux and, as discussed below, has various interpretations including diffusion and convection. In addition, as discussed below, there are interfacial effects that inhibit transport across the interface and need to be accounted. Before examining the various governing equations, it is worthwhile to discuss the underlying morphology and multiscale transport processes of the membrane as shown in Fig. 4.

Nafion and almost all PFSA ionomers have a phase-separated nanostructure. Hydration (water uptake) of the membrane leads to a nanoscale phase separation of hydrophobic PTFE region and hydrophilic sulfonic acid region. Nanophase separation and water-uptake capacity are affected by the backbone rigidity and ionic moieties and concentrations, which control the reorganization and interconnectivity of the domains. Thus, the membrane's characteristic properties are related to its phase-separated nanostructure. In addition, as mentioned, water plays a key role in the ion-transport mechanisms in PFSA membranes [34–36]. The morphology of PFSA membranes has been under investigation over the past few decades. The early work of Roche et al. [37, 38], Gierke et al. [39, 40], and Fujimara et al. [41, 42]. suggested a phase-separated nanostructure with a humidity-dependent ionomer peak due to the hydrophilic domains [37–40, 43]. Gierke and coworkers [39, 40] proposed the cluster-network model where water domains form interconnected spherical clusters of 4 to 6 nm diameter in the polymer. Other morphological descriptions for water-swollen PFSA membranes include a polymer/water layered structure [44], a disordered network of polymer chains and water [45], parallel-cylindrical water channels in semi-crystalline polymer matrix [46], and a bicontinuous network of ionic clusters in a matrix of fluorocarbon chains [47]. Today, it is still not definitively known what the microstructure seems to be, but something akin to ribbons of hydrophilic domains interspersed among backbone crystallites is the leading candidate [48, 49]. It is also known that a fluorocarbon-rich skin forms on the surface of Nafion® depending on the humidity [50–58], where studies have shown an increase in overall hydrophilicity of the surface with humidity.

Depending on the membrane's water content, the proton transport mechanism varies and has a strong effect on conductivity. The membrane conductivity is determined by size, shape and connectivity between the hydrophilic clusters. The water sorption in the membrane is classified into two groups: bound water and bulk water. Larger clusters have more bulk water and smaller clusters have more bound water and both of these have different dynamics affecting the overall conductivity. These two dynamics led to a hypothesis that there are two different proton transport mechanisms. Proton mobility across the membrane is hypothesized to occur by (1) hopping mechanism—observed at high-humidity condition and on the order of picoseconds and (2) vehicular mechanism—prominent at low-humidity condition and on the order of nanoseconds.

The key metric for membrane properties is the water content, λ (= mol H_2O/mol SO_3^-), which can be calculated from the water mass uptake of the membrane as

$$\lambda = \frac{M_w/\bar{V}_w \rho_w}{M_p/\text{EW}} \tag{46}$$

where EW is equivalent membrane [g/mol] of the membrane, $\bar{V}_w$ is the (partial) molar volume of water (~ 18 [cm^3/mol]) that may change slightly during uptake, and M_w and M_p are the mass of water and dry polymer, respectively. The water concentration in the membrane is then

$$c_w = \frac{\lambda}{\lambda \bar{V}_w + \bar{V}_p} = \frac{\lambda}{\lambda \bar{V}_w + \text{EW}/\rho_p} \tag{47}$$

where $\bar{V}_p$ and ρ_p are the molar volume and density of dry polymer, respectively. The volume fraction of water in the hydrated membrane is also commonly used, which is simply

$$\varphi_w = c_w \bar{V}_w \tag{48}$$

In most experimental setups, the controlled parameter is the water-vapor activity, a_w, (or relative humidity) instead of water content. Thus, the relationship between the water content and a_w at a given temperature, so-called sorption isotherms, must be determined. Figure 5 shows a sample of the data for such a plot.

As the membrane becomes more hydrated, the sulfonic acid sites become associated with more water, allowing for a less bound and more bulk-like water to form. This new water is no longer strongly influenced by the dielectric properties of the sulfonic acid groups and is essentially enlarging the ionic domains by filling them in

Fig. 5 Membrane water content, λ, as a function of water activity at ambient temperature from various experimental data [40, 56, 59–64]

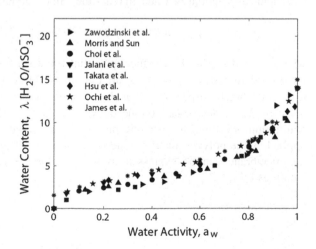

with water. This is why there is a flattening out of the slope above $\lambda = 6$ in the uptake isotherm. The extreme case is when the membrane is placed in a liquid water reservoir, where the ionic domains swell and a bulk-like liquid-water phase comes into existence throughout the membrane. In this case, there is a large increase in water uptake ($\lambda = 22$–24), and the uptake difference between 100 % relative humidity and liquid is known as Schröder's paradox since the chemical potential is the same [57, 65–67]. It is now believed that this paradox is caused by surface films and their impact on overall structure and chemical potential within the membrane.

3.1.1 Calculating Water Uptake

As water content, λ, is such a critical variable for predicting transport properties and describing the membrane state, extensive efforts have been undergone to predict λ based on measurable properties and environmental conditions. Almost all of the recent efforts resort to a thermodynamic, equilibrium-based energy-balance approach to explain the sorption phenomena based on the contributions from the elastic forces and electrostatic interactions [66, 68–77]. These models typically either assume a nanoscale morphology or calculate one. For example, the models of Freger and coworkers [69, 70] and Promislow and coworkers [64, 65], both use an expression for the free energy of the system and minimize it among possible geometries to derive a description of the hydrophilic domain microstructure. In this fashion, they can predict the impact of environment on water uptake and provide a domain distribution that can be used to run transport simulations. In all of the models, the key is that the thermodynamic equilibrium is governed by a mechanical/chemical energy balance where the sulfonic acid moieties would like to dissolve but this is hindered by the backbone, hydrophobic moieties that generate a swelling pressure [69, 75].

To understand water uptake, one starts with the fact that the swelling of a membrane at a given humidity and temperature is governed by the equilibrium of the chemical potential of water (having the same reference state)

$$\mu_w^e = RT \ln a_w = \mu_w^p = RT \ln a_p + \overline{V}_w \Pi(p_s) \tag{49}$$

where μ_w^e and μ_w^p are the chemical potential for water external and internal to the membrane, respectively, a_w and a_p are the activity of the water external and internal to the membrane, respectively, T is the absolute temperature, R is the universal gas constant, $\overline{V}_w$ is the molar volume of water, and Π is the osmotic pressure. For equilibrium swelling, the osmotic pressure must equal the swelling pressure, p_s, applied by the polymer matrix to the water domains. The water activity internal to the membrane can be expressed using the Flory–Huggins theory for polymer solutions [78],

$$\ln a_p = \ln\left(1 - \phi_p'\right) + \left[\left(1 - \frac{1}{\bar{V}_p/\bar{V}_w}\right)\phi_p' + \chi\phi_p'^2\right] \tag{50}$$

where χ is Flory–Huggins interaction parameter, which characterizes the enthalpic interactions of mixing between the polymer and solvent, and ϕ_p' is the volume fraction of the polymer including the bound water, i.e.

$$\phi_p' = \frac{\bar{V}_p + \lambda^B \bar{V}_w}{\bar{V}_p + \lambda \bar{V}_w} \tag{51}$$

where λ^B is the bound water that is strongly attached to ionic groups [64, 79, 80]. Thus, the total water content in the membrane consists of chemically bound water and free water [64, 73, 79, 81–84]. The molar volume of the dry polymer,

$$\bar{V}_p = EW/\rho_p \tag{52}$$

is related to the equivalent weight (EW) of the membrane and its dry density, ρ_p.

As the interaction parameter of a solvent/polymer network decreases, solvent uptake becomes more favorable. The interaction parameter could be determined empirically, for example by fitting the Flory–Huggins expression to experimental water-uptake data, or calculated using atomistic models. For PFSA membranes, the reported values for χ are between 0.9 and 2.5 with a strong dependence on the water content [79, 85, 86],

$$\chi(\phi_p, T) = \chi_S \phi_p^{1.5} + \chi_T\left(1 - \frac{T}{T_{\text{ref}}}\right) \tag{53}$$

where χ_S and χ_T are the components of the interaction parameter controlling the swelling and temperature effects, respectively, and $T_{\text{ref}} = 298$ K is the reference temperature. This results in Eq. 49 becoming

$$\left(1 - \phi_p'\right)\exp\left[\left(1 - \frac{1}{\bar{V}_p/\bar{V}_w}\right)\phi_p' + \chi(\phi_p, T)\phi_p'^2\right] - a_w \exp\left(-\frac{\bar{V}_w}{RT}\Pi(\phi_p, T)\right) = 0 \tag{54}$$

To solve Eq. 54, one needs to know the swelling pressure function, $\Pi(\phi_p, T)$. Several approaches exist in the literature to correlate the swelling pressure to the water volume fraction [69, 73–75]. A typical one is to use empirical information about nanoscale domains as measured by small-angle X-ray scattering (SAXS) and assuming a given geometry (e.g., cylinders). The normalized pressure generated in the network, p_s/E_{pm}, is due to the radial deformation of the backbone during swelling from the dry state,

$$\frac{\Pi(\phi_w, T)}{E_{pm}(T)} = \frac{p_s}{E_{pm}} = 1 - \frac{\Delta}{\Delta_{dry}} = 1 - \frac{1/2d - r}{1/2d_{dry} - r_{dry}} = 1 - \frac{d}{d_{dry}}\left(\frac{1 - \phi_{pore}^{1/n}}{1 - \phi_{pore}^{dry1/n}}\right)$$

$$(55)$$

where r is the radius of the hydrophilic water domains, n is the dimension of the morphology (e.g., $n = 2$ for cylindrical domains and 3 for spherical domains), $\phi_{pore} = (2r/d)^n$ from geometry, and d/d_{dry} is determined from SAXS experiments. The temperature dependence is implemented into Young's modulus of the polymer [75]. When the membrane is completely dry, it is assumed that the domains contain only the SO_3^- groups. Therefore, ϕ_{pore}^{dry} must be equal to the SO_3^- volume fraction of a dry PFSA membrane,

$$\phi_{pore}^{dry} = \phi_{SO_3}^{dry} = \frac{\bar{V}_{SO_3}}{\bar{V}_p} = \frac{\bar{V}_{SO_3}}{EW/\rho_p} \tag{56}$$

where $\bar{V}_{SO_3}$ can be taken to be 40.94 cm^3/mol [39], and the pore volume fraction becomes

$$\phi_{pore} = \phi_w + (1 - \phi_w)\phi_{pore}^{dry} \tag{57}$$

If one wants to account for other external body forces acting on the membrane (e.g., constraint or compression), this can readily be accomplished [66, 87]. Since the thermodynamic equilibrium is always maintained even with constraints on the membrane as discussed by Weber and Newman [88], equilibrium swelling of a compressed membrane can be written by modifying the pressure term in the chemical potential of water inside the polymer in Eq. 49,

$$\mu_w^{p,c} = RT \ln a_p + \bar{V}_w \Pi(p_s, p_e) \tag{58}$$

The new pressure term becomes a function of the original swelling pressure, p_s, and the applied external pressure, p_e.

Other approaches to determine the water content revolve more around treating the membrane more as akin to a porous medium where there are different types of channels related to the internal interaction energies [61, 89–92]. These more mesoscopic approaches are somewhat beyond the scope of this chapter but are worth noting as they provide a means to describe a transition from the liquid- to vapor-equilibrated structures (i.e., they bridge Schröder's paradox). They also typically use single-pore interaction equations to predict the swelling pressure using similar local equilibrium arguments as above [77, 93]. In such a scheme, one assumes there are pores within the membrane that are either liquid-equilibrated or vapor-equilibrated. These are distributed in a random network and the swelling pressure and energy of the different pores is used to predict the fraction of those pores and the overall membrane water content. In addition, this can also be used to examine transport through the membrane's mesoscale morphology.

3.2   Transport Equations

3.2.1   General Governing Equations

The overall transport and species balances described in Sect. 2 applied at the macroscale remain valid. In the membrane there is no consumption or generation of charge or species, so the source terms, S, can be neglected. Since the membrane is assumed to contain only protons as charge carriers (the transference number of protons is 1), the ionic current density in the membrane phase (subscript 2) is given by equation

$$i_2 = Fz_+N_+ = FN_+ \tag{59}$$

The simplest way to treat proton transport across membrane is to use Ohm's law, as in Eq. (3), relating the current and potential under a constant conductivity condition,

$$i_2 = -\kappa\nabla\Phi_2 \tag{60}$$

where κ is the ionic conductivity of the membrane and $\nabla\Phi_2$ is the potential gradient across the membrane (main driving force). Such a simple treatment is used in models that focus on non-membrane components. However, it is known that proton conductivity is not constant and consideration of other factors as mentioned already requires a more rigorous treatment that involves the transport of water along with that of protons.

As noted, for a rigorous treatment of the transport of proton and water, one can use either a dilute-solution resulting in Nernst–Planck Equation (25) or concentrated-solution approach. Similarly, for water, dilute-solution theory results in a Fickian type Eq. (12),

$$N_w = -\alpha\nabla\mu_w \tag{61}$$

where α is called the transport coefficient and depends on the type of driving force that is chosen as discussed later. However, it is known that the cross terms between water and proton flux are not negligible, and thus need to be accounted for. Using concentrated-solution theory, one considers the various binary interactions in the system. Doing this results in the two coupled equations:

$$i_2 = -\kappa\nabla\Phi_2 - \frac{\kappa\xi}{F}\nabla\mu_w \tag{62}$$

and

$$N_w = -\frac{\kappa\xi}{F}\nabla\Phi_2 - \left(\alpha + \frac{\kappa\xi^2}{F^2}\right)\nabla\mu_w \tag{63}$$

where ξ is the electro-osmotic coefficient and is defined as the ratio of flux of water to the flux of protons (in the absence of concentration gradients).

$$\xi = \frac{N_w}{N_+} \tag{64}$$

and the potential in the membrane can be defined by

$$\Phi_2 = \frac{\mu_+}{F} \tag{65}$$

These two equations combined with the material and charge balances and energy equation (Eqs. 10, 20, and 44, respectively) constitute a closed set of independent equations that completely describe the transport within the membrane for a concentrated-solution system composed of water, proton, and membrane.

The boundary conditions used in conjunction with the above equations can vary and are to some degree simulation dependent. Normally, the current density, water flux, reference potential, and water chemical potential are specified; but two water chemical potentials or the potential drop in the membrane can also be used. If modeling more regions than just a membrane, additional mass balances and internal boundary conditions must be specified. In particular, as it is known that the interface of the membrane can represent a mass-transport resistance, the membrane water-uptake boundary condition is altered to include a mass-transfer coefficient instead of assuming an equilibrium isotherm directly (i.e., using Fig. 5 or Eq. 54)

$$N = k_{mt}(a_{in} - a_{out}) \tag{66}$$

where in and out refer to the water activities directly inside and outside of the membrane interface and k_{mt} is a mass-transfer coefficient. This approach is the same as including a surface reaction (e.g., condensation) at the membrane interface. Finally, since the water is in a condensed state within the membrane, water uptake should involve the release or consumption of phase change heat, which should be accounted for in the energy conservation Eq. (44).

It should be noted that the above discussion and equations have not stated anything about the mode of transport; thus, the equations above are general.

3.2.2 Choice of Water Driving Force and Transport Parameters

As noted in the previous section, there are three main transport properties within the membrane: conductivity, electro-osmotic coefficient, and transport coefficient. All have been experimentally measured and are functions of temperature and water content (λ) for a given membrane. The first two are relatively straightforward to interpret and use [91]. Typically, the values are taken from data and empirical expressions are used, although there are some more mesoscopic and nanoscopic models that aim to predict these values, which is an opportunity for multiscale modeling to relate the macroscopic observables and coefficients to polymer morphology and environmental conditions. The transport coefficient requires some further discussion.

As mentioned, one can interpret the chemical-potential driving force in Eq. 61 in terms of various means, with the most popular being the use of concentration (i.e., λ) or pressure, termed diffusive and hydraulic models, respectively [18]. Increasingly, some models use both as separate driving forces, thus allowing for the determination and use of two transport coefficients [77]. This method provides an increase in the degree of freedom for solving the problem, which allows one to handle effects like Schröder's paradox; however, it is not thermodynamically rigorous. The choice of driving force informs the use of transport coefficient as well as interpretation of experimental data.

The most rigorous interpretation is to use the chemical potential directly, which is composed of activity (concentration) and pressure terms,

$$\nabla \mu_w = RT \nabla \ln a_w + \bar{V}_w \nabla p \tag{67}$$

In this fashion, the above accounts for both vapor- and liquid-equilibrated transport mechanisms using the single, thermodynamic driving force, with the transport coefficients being averaged in some fashion. However, the use of a chemical-potential driving force does become hard to determine when used in conjunction with temperature gradients due to the presence of the partial molar entropy in its definition [94].

For a pressure-driven process, the measured variable is the permeability (see Darcy's law, Eq. 36) and thus

$$\alpha_L = \frac{k_{sat}}{\mu \bar{V}_w^2} \tag{68}$$

where the L denotes that it is from a liquid-equilibrated measurement, μ is the viscosity, and k_{sat} is the membrane permeability. For an activity-driven process, the diffusivity is the measured parameters and [91]

$$\alpha_V = D_\mu \frac{c_w(\lambda + 1)}{RT} \tag{69}$$

where D_μ is the diffusivity relative to a thermodynamic driving force. While this is valid for interpretation from NMR measurements, it can be correlated to other driving forces [95]. For example, one can use thermodynamic identities showing that

$$D_f = \frac{\partial \ln a_w}{\partial \ln c_w} D_\mu \qquad (70)$$

where D_f is the diffusion coefficient for steady-state (Fickian) diffusion that is related to a gradient in water concentration. In addition, some experiments measure a dynamic diffusion coefficient during sorption or desorption, which typically has a different dependence on water content than those measured by steady-state measurements [95]. The issue is that during sorption and desorption, polymer relaxation is also occurring. In this situation, the dynamic diffusivity (although, as discussed, it is not really just a diffusive process), D_d, is determined from fitting the data using Fick's second law,

$$\frac{\mathrm{d}c_w}{\mathrm{d}t} = -\nabla \cdot N_w = -\nabla \cdot D_d(c_w)\nabla c_w \qquad (71)$$

The equation above can be solved with the initial and boundary conditions reflecting the nature of the water transport, where one uses an interfacial boundary condition of the type

$$D_d \frac{\mathrm{d}c}{\mathrm{d}z} = k_m(c_\infty - c_0) \qquad (72)$$

where k_m is the interfacial mass-transport coefficient and c_∞ is the concentration of water in the environment.

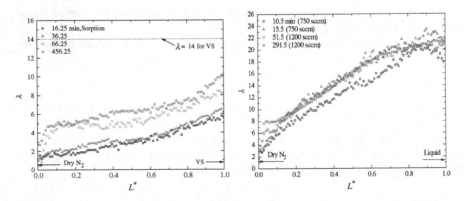

Fig. 6 Measured, time-resolved water profiles across Nafion using X-ray computed tomography and changing at $t = 0$ from dry boundary to either vapor saturated (*left*) or liquid (*right*). Figures reprinted from Ref. [96] with permission

To understand the various water transport processes and if they occur during operation, Hwang et al. measured the water profiles across a Nafion membrane [96]. As shown in Fig. 6, it is clear that there is a difference between the dynamic and steady-state profiles, including a possible change in slope that is indicative of changing diffusivity functionality. In addition, it is clear that there is interfacial resistance, which is apparent in that the vapor-saturated side does not reach close to $\lambda = 14$. Also, there is the implied nonlinear functionality of the diffusion coefficient, where the plateau around $\lambda = 6$ or so corresponds to fast Fickian diffusion which is derived from the Darken activity correction (Eq. 70) as calculated by the sorption isotherm shape (Fig. 5). Figure 6 clearly shows a much more rapid approach to steady state in liquid water as well as no interfacial resistance on the liquid side, agreeing with other studies [97–101]. The plateau at high λ is also indicative of α_L being much larger than α_V [102], with the end of the plateau perhaps indicating the change from liquid to vapor-equilibrated transport mechanisms. This observation is also in agreement with measurements that say that transport is relatively fast within the membrane and is limited by interfacial mass transfer as discussed above.

3.2.3 Gas Crossover

In addition to the water and proton species, it is important to also account for gas crossover. The crossover of these gases results in a mixed potential at the electrode —thus explaining the difference between the observed open-circuit potential and the equilibrium potential and a chemical short of the cell. Although the crossover is normally only a small efficiency loss, it does limit the thickness of the membrane [103], and can become important if pinholes or membrane thinning occur. Furthermore, crossover is attributed to carbon corrosion during fuel starvation [104], platinum-band formation [105], and peroxide generation [106]. In addition, recent studies have also shown that the dilution effect by crossover of nitrogen can be important, especially for systems that recirculate the anode stream [107, 108].

Since the gases are dilute in the membrane, it is easiest to just use experimentally measured permeation coefficients (which increase with water content and temperature)

$$\mathbf{N}_i = -\psi_i \nabla p_i \tag{73}$$

where ψ_i and p_i are the permeation coefficient and partial pressure of species i, respectively. A permeation coefficient is used instead of separate diffusion and solubility coefficients since it simplifies the analysis and the need for experimental data, and because the individual properties typically have offsetting temperature and water-content dependences.

3.3   Membrane Swelling

As discussed in determining λ (Sect. 3.1.1), there is a swelling pressure that generates macroscopic changes in the membrane's dimensions at high relative humidities. The swelling is non-affine and one can do a mechanical energy analysis to understand and predict the dimensional changes [66, 75, 76, 88, 109, 110]. Typically, one assumes additive constant molar volumes,

$$V = \bar{V}_m + \lambda \bar{V}_w \tag{74}$$

which is only rigorously true at higher (free-swelling) water contents since the initial water molecules solvate the molecules and do no swelling work as discussed above. The impact of swelling is the change in the overall volume of the membrane, which impacts the concentration and gradient magnitudes through thickness calculations when modeling. In addition, membranes are often constrained in cells, which then require analysis of Poisson's effects and areal constraint and even issues with the uniaxial compression that is applied during cell assembly. That being said, most membranes have quite high moduli, thereby meaning that they will swell normally, especially in the thickness direction, where the less strong and more porous GDLs will compress [88].

Membrane swelling can be either isotropic or anisotropic and depends strongly on if and how the membrane is mechanically reinforced or if it has substantial crosslinking. To account for swelling, one can use a variable transformation

$$Z' = \frac{z}{1 + s\frac{\bar{V}_w}{\bar{V}_m}\lambda} \tag{75}$$

where z is a given spatial dimension, s is the swelling factor in dimension z, and $\bar{V}_m$ is the partial molar volume of the dry membrane (Eq. 52). In this scheme, one calculates the swelling dimensional change based on the average water content of the membrane, $\hat{\lambda}$, using an expression like in Eq. 75, where s depends on the anisotropy of the swelling and the partial molar volumes can be a function of λ (typically they are taken to be constant as mentioned). However, since $\hat{\lambda}$ is not known a priori, one must iterate over the entire simulation until the value converges or one can use the following set of differential equations to calculate it during the simulation. The first equation is an expression of the average water content as an integral

$$\hat{\lambda} = \frac{1}{l}\int\limits_0^{z=l} \lambda(z)\mathrm{d}z = \int\limits_0^1 \lambda(\varsigma)\mathrm{d}\varsigma$$

where the equation has been nondimensionalized. The second equation arises from the thickness being a scalar quantity that is uniform

$$\frac{dl}{d\varsigma} = 0$$

These two equations are solved with the swelling boundary condition given by Eq. 75 and applied at $\varsigma = 1$. It should be noted, that recent advances with reinforced membranes have shown better mechanical properties and hence less swelling of the membrane [111]. In this case, macroscopic swelling can be ignored.

Swelling and dimensional changes can also result in mechanical failure of the membrane. This arises due to the shrinkage and contraction of the membrane during operation where the local humidity changes [112, 113]. Often this mechanical durability is related to chemical degradation as well [114]. This latter form of degradation is related to chemical attack [113, 115, 116] and is often simulated using more molecular simulations [117]. The area of durability provides a good future opportunity for multiscaling since durability and degradation issues are typically considered only as changing the macroscopic parameters of the cell with time, and more information into the actual mechanisms should be linked with macroscopic performance models. In this fashion, the various stressors can be evaluated and their impact predicted in terms of lifetime and performance.

3.4 Contamination and Multi-ion Transport

The analysis and discussion above is centered on having the proton being the only mobile ion inside of the membrane. There are instances where this is not the case, for example, that of cation contamination including Pt ions due to Pt dissolution, flow-field corrosion products, salts from the environment, etc. [20, 113]. These ions will ion-exchange with the protons in the membrane and can cause dramatic decreases in cell performance. In addition, the issue of multi-ion transport in the membrane is especially critical in alkaline-exchange membranes, where carbon dioxide from the air will cause a competition between hydroxide and bicarbonate ions in the membrane.

To calculate the distribution of ions in the membrane in contact with a reservoir, one needs to consider chemical equilibrium among the various species in the membrane and their counterparts in the external reservoir. To do this, one needs to calculate the concentration distribution of each ion inside of the pore. This distribution can be written as a modified Boltzmann distribution (see Eq. 22)

$$c_i(r) = c_i^{ext} \exp\left(-\frac{z_i F \Phi(r)}{RT} - \frac{A_i}{RT}\left(\frac{1}{\varepsilon_r(r)} - \frac{1}{\varepsilon_r^{ext}}\right)\right) \tag{76}$$

where r is the radial position of the pore, ext denotes the external reservoir, A_i is an ion hydration constant of species i (see Ref. [118] for values), and ε_r is the dielectric of the medium. In the above expression, the first term represents electrostatic attraction/repulsions of ions and the second relates the effects of a changing dielectric medium. If the dielectric is constant and the same as the reservoir, then the expression becomes a normal Boltzmann distribution.

To calculate the distribution, the dielectric constant and potential distributions must be known. The dielectric constant distribution can be determined using Booth's equation [119]

$$\varepsilon(r) = n^2 + \frac{3\left(\varepsilon_r^{ext} - n^2\right)}{\zeta\nabla\Phi(r)}\left[\coth(\zeta\nabla\Phi(r)) - \frac{1}{\zeta\nabla\Phi(r)}\right] \tag{77}$$

where n is the refractive index of the solution and ζ is

$$\zeta = \left(\frac{5\eta}{2k_BT}\right)(n^2 + 2) \tag{78}$$

where k_B is Boltzmann's constant, and η is the dipole moment of the solvent molecule. Finally, the potential distribution can be calculated by solving Poisson's Eq. (21). Solving Eqs. (21), (76) and (77) simultaneously yields the concentration distributions of the various ions and the potential distribution within the pore.

The results of the partition calculations demonstrate an order of magnitude larger dielectric constant near the pore walls where the sulfonic-acid sites reside than in the pore middle [36, 120]. In fact, the change of the dielectric constant can be correlated to the existence of bulk-like water. The above approach allows the prediction of ion partitioning by fuel-cell membranes [27]. For alkaline membranes, one must also consider the equilibrium between carbon dioxide and its carbonate forms.

The above analysis yields distributions of the potential and ions, but does not treat their transport. As discussed above, Nernst–Planck Eq. (25) can be used for the transport along with the multi-ion definitions of current density (Eq. 18) and conductivity (Eq. 28), the modified Ohm's law (Eq. 27), and electroneutrality (Eq. 11). Due to the presence of more species, more properties are required. These can take the form of the conductivity, and species' mobilities, diffusion coefficients, and transference numbers

$$t_j = \frac{z_j^2 u_j c_j}{\sum_i z_i^2 u_i c_i} \tag{79}$$

which is the fraction of current carried by ion j in the absence of concentration gradients. Without other ions, this value is 1 for protons, and it can be used as an indication of the current-transport efficiency. Also, the electro-osmotic coefficient is basically the transference number of water (see Eq. 64).

However, for these multicomponent systems, concentrated-solution theory is more appropriate as it contains the correct number of transport properties and also the binary ion/ion interactions are expected to be important. The downside is that the analysis is much more complex and requires more knowledge of the various transport properties and activity coefficients. For this analysis, equations of the form of Eq. 31 can be used along with the definitions and electroneutrality. For example, for a four-component system composed of protons (H^+), single-charged cations (X^+), water (w), and membrane (M^-), Eq. 31 can be written as

$$\nabla \mu_{X^+} = RT \left(\frac{\nabla x_{X^+}}{x_{X^+}} - \frac{\nabla x_{H^+}}{x_{H^+}} \right) + b \left(\nabla \left(y_{HM}^2 \right) - \nabla \left(y_{KM}^2 \right) \right) + F \nabla \Phi_2 \tag{80}$$

$$\nabla \mu_{M^-} = RT \left(\frac{\nabla x_{H^+}}{x_{H^+}} + \frac{\nabla x_{M^-}}{x_{M^-}} \right) + b \nabla \left(y_{XM}^2 \right) - F \nabla \Phi_2 \tag{81}$$

and

$$\nabla \mu_w = RT \frac{\nabla x_w}{x_w} \tag{82}$$

where the potential was defined relative the hydrogen potential as before (see Eq. 65), y_i is the mole fraction of cation i relative to the total number of cations, and b is a constant related to the activity coefficient. These equations can be solved along with the concentration-dependent water-uptake isotherm, diffusivities, and activity coefficients and the equations discussed above.

Finally, some of the pore models utilize the above methodology and equation set along with the thermodynamic descriptions and swelling pressures to predict water uptake and ion transport [28, 36, 121–123]. These mesoscale models are then scaled through statistical distributions up to macroscopic realms where the entire membrane is simulated. Such analysis is beyond the scope of this chapter, but it

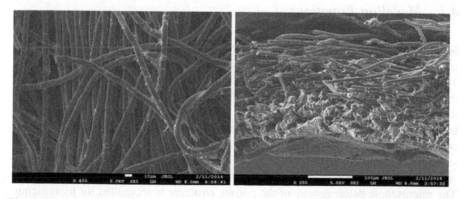

Fig. 7 Scanning electron micrograph of (*left*) GDL surface and (*right*) GDM cross-section with microporous layer on the bottom and GDL on the top

does provide a way in which continuum modeling can provide insight at multiple scales.

4 Gas-Diffusion Media

Gas-diffusion media (GDM) or porous-transport layers (PTLs) are composite structures typically comprised of a macroporous gas-diffusion layer (GDL), usually made of woven carbon cloth or carbon paper, and a microporous layer as shown in Fig. 7. They sit between the gas channels and the catalyst layers, with the catalyst layers facing the microporous layer. They provide pathways to disperse reactant fuel, oxidant, heat, and electrons while removing product water. A GDM has to be hydrophilic enough to wick out the water and hydrophobic enough to not fill with liquid water and "flood" and block the reactant gas from reaching the catalyst site. This seemingly competing objective is met by partial treatment of the naturally mixed wettable layers with hydrophobic PTFE. Water produced at the cathode and water transported across the membrane is removed out of the cell by capillary effects including perhaps through cracks and preferential pathways. A microporous layer serves to protect the membrane from being penetrated by the carbon fibers of the macroporous GDL, provide discrete locations of water injection into the GDL, and decrease interfacial roughness or porosity. This decreases the water accumulation near the cathode and hence decreases mass-transport resistance to oxygen diffusion. In this section, we discuss the mathematical description of GDM along with the assorted phenomena mentioned above. Because of similarity, all the transport equations described are applicable to the description of the catalyst layers that will be discussed in Sect. 5.

4.1 Modeling Equations

Modeling the GDM involves descriptions of the fluxes in the gas and liquid phases, interrelationships among those phases, as well as electron and heat transport. Traditional equations including Stefan–Maxwell diffusion (Eq. 15), Darcy's law for momentum (Eq. 36), and Ohm's law for electron conduction (Eq. 29) are typically used in macroscale, macrohomogeneous situations, where most of the effective transport properties of the various layers have been measured experimentally [124], or perhaps modeled by more microscopic methods. Effective properties are required since the continuum approach treats the material as continuously distributed in space, but the porous media contain multiple phases due to solid fibers or particles. The microscopic heterogeneity of the porous structure is accounted for by utilizing effective properties. This is accomplished by volume averaging all the relevant properties and system variables for transport within the porous domain,

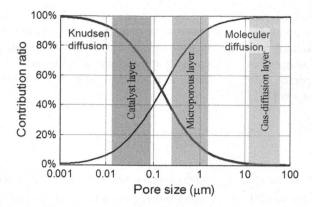

Fig. 8 Relationship between domain pore size and the contribution ratio of molecular diffusion versus Knudsen diffusion of oxygen in nitrogen at 80 °C, 150 kPa. Reprinted from Ref. [125] with permission

$$\psi_k^{\text{eff}} = \frac{\varepsilon_k}{\tau_k} \psi_k \tag{83}$$

where ψ_k represents any property in the phase k and ε_k and τ_k are the porosity and tortuosity of phase k, respectively. Of course, there are always caveats and simplifications that must be made to model the macroscale transport. Thus, the specific pore structure is often considered only in a statistical sense, local equilibrium among phases is often assumed, and the effective properties, which are often measured for the entire layer, are applied locally and assumed to remain valid. Finally, only recently have issues related to specific interfacial phenomena, contact resistances, and correlating the channel conditions to water droplets and removal been studied and included, even though such effects could dominate the overall response of the cell. For example, cell compression and solid mechanics can be used to understand how assembly may change the effective properties including lower porosities in the GDL and perhaps even bowing of the GDL into the flow-field channel; such studies are beyond the scope of this chapter.

4.1.1 Gas Phase

Treatment of the gas-phase species is relatively straightforward where there could be two main diffusion mechanisms as seen in Fig. 8. From an order-of-magnitude analysis, when the mean-free path of a molecule is less than 0.01 times the pore radius, bulk diffusion dominates, and when it is greater than 10 times the pore radius, Knudsen diffusion dominates. This means that Knudsen diffusion is significant when the pore radius is less than about 0.5 μm. For reference, a typical carbon gas-diffusion layer has pores between 0.5 and 20 μm [126–128] in radius, and a microporous layer contains pores between 0.05 and 2 μm [129, 130]. Thus, while Knudsen diffusion may not have to be considered for gas-diffusion layers, it should be accounted for in microporous and catalyst layers. Such transport can be considered as interactions between the gas molecules and the wall. If the wall is taken as a species, one can derive a modified Stefan–Maxwell equation for gas-phase diffusion [131, 132]

$$\nabla x_i = -\frac{N_i}{c_T D_{K_i}^{\text{eff}}} + \sum_{j \neq i} \frac{x_i N_j - x_j N_i}{c_T D_{i,j}^{\text{eff}}}$$

where the $D_{K_i}^{\text{eff}}$ is the effective Knudsen diffusion coefficient. The effective diffusivities are both a function of the bulk porosity and the saturation, S, also known as the liquid volume fraction of the pore space,

$$D_{i,j}^{\text{eff}} = \frac{\varepsilon_G}{\tau_G} D_{i,j} = \frac{\varepsilon_0(1-S)}{\tau_G} D_{i,j} \tag{84}$$

where τ_G is the tortuosity of the gas phase and ε_0 is the porosity of the medium. While a Bruggeman expression is often used for the tortuosity effect,

$$\tau_G = \varepsilon_G^{-0.5} \tag{85}$$

This has been shown not to be valid for fibrous GDLs, where the exponential power has been measured to be closer to 2–3 for each effect [133–135],

$$\frac{D_{i,j}^{\text{eff}}}{D_{i,j}} = \varepsilon_0^{3.6}(1-S)^3 \tag{86}$$

The above also agrees with more microscopic modeling results [136, 137]. For the bulk movement and convection of the gas phase, Darcy's law and the mass-averaged velocity (Eqs. 36 and 13, respectively) are used with an effective permeability that is comprised of the absolute (measured) permeability and a relatively permeability owing to the impact of liquid

$$k_k = k_{r,k} k_{\text{sat}} \tag{87}$$

where $k_{r,k}$ is the relative permeability for phase k. The above equations are used along with the transient mass balance (Eq. 10) to describe the gas-phase transport.

4.1.2 Liquid Phase

While one can treat the liquid water as a mist or fog flow (i.e., it has a defined volume fraction but moves with the same superficial velocity of the gas), it is more appropriate to use a separate equations for the liquid. This is often of the form of Darcy's law (Eq. 36), which in flux form is

$$\mathbf{N}_{w,L} = -\frac{k_L}{\overline{V}_w \mu} \nabla p_L \tag{88}$$

where $\overline{V}_w$ is the molar volume of water. One can also add a second derivative to the above such that a no-slip condition can be met at the pore surfaces (i.e., Brinkman equation) [138]. There are also modeling methodologies that reformulate the transport equation. For example, one can use the saturation as the driving force, resulting in a governing equation of

$$\mathbf{N}_{w,L} = -D_S \nabla S \tag{89}$$

where D_S is a so-called capillary diffusivity

$$D_S = \frac{k}{\mu \overline{V}_w} \frac{dp_C}{dS} \tag{90}$$

Although the above equation is valid, it gives the false impression that the saturation is the driving force for fluid flow, and that a saturation condition should be used as a boundary condition. Furthermore, care must be taken in the interpretation of the capillary diffusivity.

4.1.3 Heat Transport

For GDM, the thermal balance turns mainly into heat conduction due to the high thermal conductivity compared to convective fluxes. Although no electrochemical reactions are occurring within the GDM, there are still phase-change reactions that can consume/generate a considerable amount of heat. Thus, those source terms must be included in the overall heat balance (Eq. 44), and it should be noted that effective properties are again required to be used in the governing equation. Finally, the contact resistance at the boundary can be appreciable for both thermal and electrical contact. This boundary condition is similar to an interfacial resistance

$$Q = h(T_{\text{in}} - T_{\text{out}}) \tag{91}$$

where the out is typically the flow field. One should note that the heat and electrical conduction typically travels through the same solid fraction and out of the cell sandwich through the ribs. Thus, they have similar property dependences including a significant anisotropy with the in-plane direction being almost an order of magnitude higher than the through-plane direction due to fiber alignment in typical GDLs [139–141].

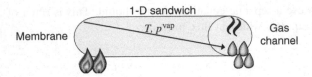

Fig. 9 Schematic of phase-change-induced flow where water and heat are transported along the cathode PEFC temperature gradient

4.1.4 Liquid/Vapor/Heat Interactions

Equations 84, 87, and 88 clearly show that there is an impact of the liquid- and gas-phase volume fractions on the transport of each other through the various effective transport properties. From a continuum perspective, these are related through the capillary pressure [30, 31, 142, 143],

$$p_C = p_L - p_G = -\frac{2\gamma\cos\theta}{r} \tag{92}$$

where γ is the surface tension of water, r is the pore radius, and θ is the internal contact angle that a drop of water forms with a solid. The functional form of the saturation dependence on capillary pressure can be measured [144, 145] or derived using various simplistic models [146] or more complicated pore-network and other models [147–150]. However, macroscopic models typically utilize microscopic-model or experimental results that are taken from analysis of the entire GDL and apply them locally; validation of this assumption still remains an open question, especially since it is known that GDM structures are not spatially homogeneous. It should also be noted that the often used Leverett J-function [136, 143] for the capillary pressure—saturation relationship was derived from hydrophilic soil systems and is not really valid for GDLs.

   Although the liquid and gas phases are related through transport properties, they also have an effect on each other's fluxes through heat transport and phase-change-induced (PCI) flow [151–153]. In this fashion, the liquid water is near equilibrated with water vapor and the temperature distribution induces a water vapor pressure gradient. The water is transported along that gradient and condenses and gives off heat at the gas channel or cooler flow field rib as shown in Fig. 9. Such an effect can be shown to be able to move all the produced water when operating above temperatures of 60 °C or so with typical component properties [152]. In this fashion, the produced water is removed in the vapor phase and flooding concerns are minimal. Such a mechanism results in substantial heat removal from the hotter catalyst layer as well. Finally, liquid water can also impact other properties such as thermal conductivity [139, 140, 154, 155].

4.2 Microporous Layers and Pore-Network Modeling

Microporous layers can be modeled in the same fashion as the GDLs discussed above. However, due to the morphology (see Fig. 7), their properties are more isotropic. In addition, these materials are typically more hydrophobic, and due to their relative thinness, interfacial phenomena are more important. Due to this effect and because they contain smaller pores, more microscopic modeling methodologies are required. This is especially true as it is believed that liquid water transports across this layer mainly through cracks and larger imperfections. The end result is that the microporous layer selectively allows water to enter the GDL, thereby necessitating a more microscopic treatment of the GDL governing phenomena since a macrohomogeneous approach is poor when such spatially significant phenomena occur. To account for such effects, pore-network modeling and other methods are required.

A pore-network model utilizes a simplified description of the pore space within the porous medium. Thus, one idealizes the geometry in terms of pores and interconnections (nodes) [147, 148]. The generated network is validated by comparison of calculated and measured parameters including the pore- and throat-size distribution data as well as measurements such as the capillary pressure—saturation relationship. Water flow and distribution within the generated network is solved by a stepwise fashion from one point to another, and thus is independent of the real-space discretization grid (i.e., it only depends on the network). For modeling transport, the same governing equations and phenomena described above hold. For example, for liquid-water imbibition, the model examines at each intersection or node where the water travels based on the local pressure and pore properties. The volumetric flowrate of water in a cylindrical pore of radius r_{ij} between nodes i and j is governed by Poiseuille flow

$$q_{w,\text{pore}} = A_{\text{pore}} \cdot v_{w,\text{pore}} = \frac{\pi r_{ij}^4}{8 \mu_{ij}^{\text{eff}} l} (\Delta p_{ij} - p_{C_{ij}}) \tag{93}$$

where Δp_{ij} is the pressure acting across the pore, l is the pore length, and $p_{C_{ij}}$ is the capillary pressure in the pore when multiple phases are present. The volumetric flowrate exists only when $\Delta p_{ij} > p_{C_{ij}}$. The effective viscosity within a pore, μ_{ij}^{eff} is a function of the fluid position inside the pore, x_{ij}, the non-wetting (injected) fluid viscosity, μ_{nw}, and the wetting (displaced) fluid viscosity, μ_w. The capillary pressure is calculated similar to Eq. 92 but where either the average radius of the intersecting pores at each node or the radius of a given pore is used depending on where the water meniscus exists,

$$p_{C_{ij}} = \gamma \cos \theta \left[\left(1 - \frac{r_i}{2\bar{r}_i} - \frac{r_j}{2\bar{r}_j} \right) \frac{1 - \cos\left(\frac{2\pi x_{ij}}{l}\right)}{r_{ij}} + \frac{1 + \cos\left(\frac{\pi x_{ij}}{l}\right)}{\bar{r}_i} + \frac{1 - \cos\left(\frac{\pi x_{ij}}{l}\right)}{\bar{r}_j} \right]$$

$$\tag{94}$$

where $\bar{r}_i$ and $\bar{r}_j$ are the average pore radius around node i and j, respectively, The capillary pressure is zero when the pore is filled with only one fluid.

Conservation of mass requires that the flowrate balance at each node for every simulation step, thus from Eq. 93 one gets

$$\sum_j \frac{r_{ij}^4}{\mu_{ij}^{eff}}(\Delta p_{ij} - p_{C_{ij}}) = 0 \tag{95}$$

where the summation is over all of the pores connecting to the node (normally 4). The unknown pressure gradient, Δp_{ij}, is solved through the equation above. In addition to the pore sizes and lengths, one also needs the pore contact angle and fluid properties. Recent advancements to the pore-network modeling include considering simultaneous heat and electron conduction through the network as well as phase change.

While pore-network models utilize a statistical representation of the actual porous medium, the most robust simulation is to use information of the pore structure and enact a direct numerical simulation of the transport equations through it [156]. In this fashion, one uses Navier–Stokes applied to the pore domain and not the more macroscopic equations like Darcy's law. Needless to say, these simulations are very computationally expensive. Similar simulations use Lattice–Boltzmann methods (LBM), which are advantageous over conventional macrohomogeneous models because of their ability to handle complex boundaries at a microscopic level [157–161]. Though LBM requires extensive computation, detailed knowledge of the various forces, and can have issues with phase-change effects and discrete phases, there is some promise in terms of determining effective properties. A competent approach then, to investigate the phenomena in their full intricacy, would be to understand and capture the physics at the microscale by combining advanced imaging (e.g., high-resolution synchrotron X-ray tomography) and levelset LBM of the water and gas transport, and then upscale it to macroscale with a continuum model validated with neutron radiography.

4.3 Transport in the Gas Channel

Similar to the GDM, the flow-field structure provides mechanical support, removal of heat and product water, efficient ingress of reactant gases, and provides a pathway for electrons. To model these effects, one can use the governing equations discussed in Sect. 2, as the transport phenomena are essentially the same as traditional fluid flow. The major issue with the channel design is to remove effectively the liquid water, especially at lower temperatures. Convection in the flow channel dominates the diffusion mode of transport, so the pressure difference drives fluid flow in the channel. The pressure drop is caused by the friction between the fluid and channel walls and laminar (i.e., Poiseuille) flow (Eq. 93) is typically assumed. In the case of stacks where gas is delivered to multiple cells at fixed pressure drop, accumulation of water droplets may lead to disparity in fuel and oxygen distribution between individual cells. For stacks operating at fixed current, this may lead to parasitic reactions that may eventually result in cell and stack failure.

While many of the 1-D and 2-D models incorporate the along-the-channel dimensional effects, such models focus only on the distribution of oxygen, fuel, etc., which essentially become boundary conditions for the more detailed 1- and 2-D simulations. For this reason, 3-D models are formulated, which try to understand the dynamics of liquid water along the flow channel. However, while this is a step in developing a durable fuel-cell stack, it does typically necessitate a reduced granularity for the individual cell layers. We believe that the optimal design in terms of computational expense and resolution need is 1 + 2-D where the rib/channel and cell dimensions are modeled rigorously, and the along-the-channel dimension is considered only at the boundaries. On the larger scale, design of suitable gas distributors and manifolds and gas channels for efficient removal of water is improved by proper modeling approaches.

The most important question that the channel model tries to address is the accumulation and transport of water droplets. This finds particular significance if one of the cells in a stack is blocked by droplets that could lead to oxygen deficiency and failure of the stack. Liquid water can exist within the gas channel as droplets or mist being carried along with the gas phase, (annular) films that flow in the corners and along the flow field, and slugs which block the channel and must be forced out. These water mechanisms can be seen as a progression, where the blockage and slug flow occur as the film and droplets agglomerate due to liquid-water buildup. To model the water flow, the multiphase approaches described above can be used where the descriptions hold most for mist flows. For film and corner flows, one can model that as flow along a parallel plate [29]. For slug flow, the system becomes inherently dynamic where the pressure builds up and moves a slug, and then water accumulation or a surface (e.g., bend in a serpentine flow-field channel) stops the slug until the pressure can increase again.

Fig. 10 Schematic of droplet force balance

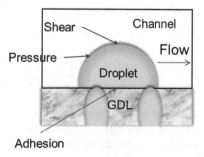

4.3.1 Droplet Movement

The above discussion centers on water that is in or condenses in the gas channel, but a major concern is removing liquid water from the surface of the GDL and into the channel. Essentially, this is a critical interfacial effect in that if droplet detachment is hindered, this creates a pressurization of the liquid in the PEFC porous layers that can lead to flooding. Determination of the correct boundary condition then is a remaining challenge for PEFC modeling. Such a condition should provide the liquid pressure and surface coverage of the droplets at a minimum.

To understand the process of detachment, a force–balance approach can be used as shown schematically in Fig. 10,

$$F_a + F_p + F_s + F_g = 0 \tag{96}$$

where F_a is the adhesion or surface tension force, F_p is the pressure force, F_s is the shear force acting on the droplet, and F_g is the gravitational force, which is negligible for typical droplet sizes although can be appreciable for large slugs. The adhesion force is best determined experimentally using a sliding-angle technique with water penetration through the bottom of the GDL as this is more accurate than the use of a contact-angle-hysteresis type of measurement on the GDL surface [162].

The shear force can be calculated based on Stokes flow past a sphere and the shape of the droplet,

$$F_s = \left[\frac{6\mu H \langle v \rangle}{(H - h)^2} \right] d^2 \tag{97}$$

where d is the droplet diameter at its maximum, H is the channel height, $\langle v \rangle$ is the average flow velocity in the channel under laminar flow [29], μ is the fluid viscosity, and h is the droplet height. Similarly, for the pressure force one can derive and expression of the form

$$F_p = \left[\frac{a\mu d \langle v \rangle}{(H - h)^2} \right] [Hd] \tag{98}$$

where

$$a = 12 \left[1 - \frac{192H}{\pi^5 W} \tanh\left(\frac{\pi W}{2H} \right) \right]^{-1} \tag{99}$$

where W is the channel width, and the above is fit from CFD simulations. The pressure force can also be used in a force balance on a water slug to determine when it will move. While the above equations can be used, rigorously, they are valid only for a single droplet in the center of a channel. For multiple droplets, interactions between them can occur and more research is required. Similarly, the impact of the underlying water-network topology on the droplets and adhesion force is not fully known; it is an area where multiscaling techniques would prove beneficial.

5 Catalyst Layer

The catalyst layer is befittingly referred to as the pacemaker of a PEFC [163]. The catalyst layers determine the rate of current generation by the electrochemical reaction. While the anode catalyst layer garnered some of the attention during the earlier part of PEFC research, due to focus on carbon monoxide poisoning of the anode, especially with reformate feeds, much of the focus is on the cathode catalyst layer due to high kinetic loss experienced by the oxygen reduction reaction. The high kinetic loss requires the use of high precious-metal (e.g., Pt) loading at the cathode. The major focus of research in this area is to increase the active area for reaction without any significant effect on the mass transport of reactant and product species. Understanding how structure, composition, and operating conditions control rates of electrocatalytic processes and their complex spatial distributions constitutes the basic task of catalyst-layer modeling. Modeling is well suited since experimental interrogations into catalyst layers are quite difficult due to the complex morphology and multiple components and phases in the layer (e.g., micro- and nanometer pores, nanoparticles of catalyst supported on slightly larger particles of carbon, thin films of ionomer, liquid films, etc.) and its overall thinness (order of 10 μm or less). The subsisting challenges and recent advances in the major areas of theoretical catalyst-layer research include: (i) structure and reactivity of catalyst nanoparticles, (ii) self-organization phenomena in catalyst layers at the mesoscopic scale, (iii) effectiveness of current conversion in agglomerates of carbon/Pt, and (iv) interplay of porous structure, liquid–water formation, and performance at the macroscopic scale. The success of any catalyst rests on the practicality of fabricating it as an electrode for PEFC operation.

Construction of catalyst layers is still an art form in which modeling always plays the catchup role (i.e., it is not predictive). For example, the cost of catalyst is reduced by minimizing the parasitic losses, increasing the power density, and decreasing the

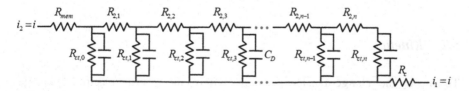

Fig. 11 Simple equivalent-circuit representation of a porous electrode. The total current density, i, flows through the separator or membrane to the electrolyte phase (2) and then into the solid or electronic phase (1) including a contact resistance. In between, the current is apportioned based on the resistances in each phase and the charge-transfer resistances, which also contain a capacitance

Pt loading. However, decreasing the catalyst loading can result in unexpected mass-transfer losses [164], as well as a sort of capacity fade due to loss of active sites to gradual processes like catalyst agglomeration (Ostwald ripening), sintering and Pt dissolution [113, 165, 166]. It is known that in the PEFC catalyst layer, "optimization has often proven to be an empirical process rather than an engineering design process largely because it is difficult to envision the ideal catalyst layer when the structure of the state-of-the-art is so poorly understood" [167]. However recent advances in molecular dynamics, mesoscale modeling, and multiscale, multicomponent coupling are showing or providing theoretical frameworks on why some catalyst layers exhibit better water and thermal management, as well as the impact of degradation [168–171]. Understanding catalyst layers is one of the most critical areas for multiscale simulations, especially as transport and related phenomena occur on disparate length scales throughout the layer and the components within the layer. Such models are described in detail in other chapters of this book; here we restrict ourselves to the more macroscopic, continuum modeling approaches. Again, the lower scales can typically provide the effective transport properties of the catalyst layer (which is even more critical than in GDM where diagnostic experiments are easier to conduct), as well as correlate those properties to structures.

From the continuum approach, one is trying to derive and use expressions for the transport of the relevant species in the various phases. One can think of something similar to Fig. 11, where the ionic current is carried by the ionomer thin films or even water layers, the electronic current is essentially shorted due to the high conductivity of the percolated carbon particles, and the charge-transfer resistance is given by the kinetic expressions. Obviously, such a picture is simplified greatly, but it does provide a means for understanding how current is distributed in porous electrodes and provides an underlying conceptual framework [27]. It should be noted that oxygen transport can also impact the current distribution, which will change the charge-transfer resistance. In addition to the more traditional macroscopic modeling of catalyst layers, some continuum models examine transport through single pores using complex reaction expressions and accounting for Poisson's equation (Eq. 21) and the space-charge region. Such models are also good at being able to analyze interesting phenomena such as ultra-thin catalyst layers that do not contain ionomer [172, 173]. Although they use a continuum Poisson–Nernst–Planck formulation (Eqs. 10, 21, and 25), they are more mesoscopic in origin and beyond the scope of this chapter.

5.1 Kinetics

To model the charge-transfer reactions within the porous electrode, kinetic expressions are used. A typical electrochemical reaction can be expressed as

$$\sum_k \sum_i s_{i,k,h} M_i^{z_i} \rightarrow n_h e^-$$ (100)

where $s_{i,k,h}$ is the stoichiometric coefficient of species i residing in phase k and participating in electron-transfer reaction h, n_h is the number of electrons transferred in reaction h, and $M_i^{z_i}$ represents the chemical formula of i having valence z_i.

The ion-transfer rate is equal to the electrochemical reaction rate at the electrodes (which is the source term, or transfer current density in Eq. 2). According to Faraday's law, the flux or species i in phase k and rate of reaction h is related to the current as

$$N_{i,k} = \sum_h r_{h,i,k} = \sum_h s_{i,k,h} \frac{i_h}{n_h F}$$ (101)

where i_h refers to the transfer current density, i.e. current (i) per unit geometric area of the electrode. The rate of a chemical reaction is related to its concentration and temperature through an Arrhenius relationship,

$$r_h = k \exp\left(\frac{-E_a}{RT}\right) \prod_i \left(\frac{a_i}{a_i^{\text{ref}}}\right)^{m_i}$$ (102)

where k is the rate constant, m_i is the order of reaction for species i, and a_i is the activity of reactant i, which as discussed above requires that an appropriate reference ion is chosen as the activity of a single ion is undefined [27]. For modeling purposes, especially with the multi-electron transport of species, it is often easiest to use a semi-empirical equation to describe the reaction rate, namely, the Butler-Volmer equation [27, 174],

$$i_h = i_{0_h} \left[\exp\left(\frac{\alpha_a F}{RT}(\Phi_k - \Phi_p - U_h^{\text{ref}})\right) \prod_i \left(\frac{a_i}{a_i^{\text{ref}}}\right)^{s_{i,k,h}}_a - \exp\left(\frac{-\alpha_c F}{RT}(\Phi_k - \Phi_p - U_h^{\text{ref}})\right) \prod_i \left(\frac{a_i}{a_i^{\text{ref}}}\right)^{-s_{i,k,h}}_c \right]$$ (103)

where i_h is the transfer current between phases k and p due to electron-transfer reaction h, the products are over the anodic and cathodic reaction species, respectively, α_a and α_c are the anodic and cathodic transfer coefficients, respectively, and i_{0_h} and U_h^{ref} are the exchange current density per unit catalyst area and the potential of reaction h evaluated at the reference conditions and the operating temperature, respectively.

In the above expression, the composition-dependent part of the exchange current density is explicitly written, with the multiplication over those species in participating in the anodic or cathodic direction. The reference potential is determined by thermodynamics as described elsewhere [27], and can commonly be determined using a Nernst equation,

$$U = U^\theta - \frac{RT}{z_i F} \ln\left(\prod a_i^{s_i}\right)$$ (104)

where s_i is the stoichiometry of species i; and the activity of the species is often approximated by its local concentration of the species. If the reference conditions are the same as the standard conditions, then U^{ref} has the same numerical value as U^θ.

The term in parentheses in Eq. 103 can be written in terms of an electrode overpotential

$$\eta_h = \Phi_k - \Phi_p - U_h^{\text{ref}}$$ (105)

If the reference electrode is exposed to the conditions at the reaction site, then a surface or kinetic overpotential can be defined

$$\eta_{s_h} = \Phi_k - \Phi_p - U_h$$ (106)

where U_h is the reversible potential of reaction h. The surface overpotential is the overpotential that directly influences the reaction rate across the interface. Comparing Eqs. 29 and 30, one can see that the electrode overpotential contains both a concentration and a surface overpotential for the reaction.

For the hydrogen oxidation reaction (HOR) occurring at the porous anode catalyst layer, Eq. 103 can be written as

$$i_{\text{HOR}} = i_{0_{\text{HOR}}}\left[\frac{p_{\text{H}_2}}{p_{\text{H}_2}^{\text{ref}}}\exp\left(\frac{\alpha_a F}{RT}(\eta_{\text{HOR}})\right) - \left(\frac{a_{\text{HM}}}{a_{\text{HM}}^{\text{ref}}}\right)^2 \exp\left(\frac{-\alpha_c F}{RT}(\eta_{\text{HOR}})\right)\right]$$ (107)

where 1 and 2 denote the electron- and proton-conducting phases, respectively, and the reaction is almost always taken to be first order in hydrogen. Typically, the dependence on the activity of the proton(H)-membrane(M) complex is not shown since the electrolyte is a polymer of defined acid concentration (i.e., $a_{\text{HM}} = a_{\text{HM}}^{\text{ref}}$). However, if one deals with contaminant ions, then Eq. 107 should be used as written. Also, it has recently been shown that the HOR may proceed with a different mechanism at low hydrogen concentrations; in this case, the kinetic equation is altered through the use of a surface adsorption term [175]. Due to the choice of reference electrode, the reference potential and reversible potential are both equal to zero.

If the system is at equilibrium then the rate of the forward reaction is equal to the rate of reverse reaction, i.e. the net current is zero and the Nernst Eq. (104) is obtained for an elementary reaction. The reaction rates at this equilibrium are written as a current density which is called the exchange current density and is defined as

$$i_{0_{\text{HOR}}} = i_{0_{\text{HOR}}}^{\text{ref}} AL \exp\left(\frac{-E_a}{RT}\left[1 - \frac{T}{T^{\text{ref}}}\right]\right) \tag{108}$$

where $i_{0_{\text{HOR}}}^{\text{ref}}$ refers to the exchange current density at the reference conditions and is based on the catalyst surface area (e.g., platinum). To change this from per unit catalyst area to geometric area, a roughness factor, AL, is used (i.e., $\text{cm}_{\text{geo}}^2/\text{cm}_{\text{Pt}}^2$). The exchange current density of a reaction is an indicator of the ease of the reaction. The exchange current density for the HOR reaction depends on catalyst being used and is very high ($\sim 1 \text{ mA/cm}^2$) for platinum, which is the typical catalyst of choice [176]. Also, it can depend on the particle size and crystal facet [163], both of which complicate continuum modeling, which typically assumes a uniform distribution and particle size. In theory, one can integrate over such distributions, but due to the complex interactions, it is often more advisable to do true multiscale simulations, where particle-size effects and distributions at the mesoscale can be handled more rigorously.

Unlike the facile HOR, the oxygen reduction reaction (ORR) is slow. Due to its sluggishness, usually for describing a cell in operation, the anodic part of the ORR is considered negligible and is dropped, resulting in the so-called Tafel approximation

$$i_{\text{ORR}} = -i_{0_{\text{ORR}}} \left(\frac{p_{O_2}}{p_{O_2}^{\text{ref}}}\right)^{m_0} \left(\frac{a_{\text{HM}}}{a_{\text{HM}}^{\text{ref}}}\right)^2 \exp\left(\frac{-\alpha_c F}{RT}(\eta_{\text{ORR}})\right) \tag{109}$$

with a dependence on oxygen partial pressure, m_0, of between 0.8 and 1 [177–180] and the same Arrhenius temperature dependence as seen in Eq. 108. For both the HOR and ORR, α is typically taken to be equal to 1 [176 178, 181, 182], however newer models use a value much closer to 0.5 for the ORR due to Pt-oxide formation [183].

The four electron ORR involves oxide formation, which form at the potential range of the ORR (0.6–1.0 V) by water or gas-phase oxygen. These oxides can inhibit the ORR by blocking active Pt sites with chemisorbed surface oxygen. Typically, a constant Tafel slope for the ORR kinetics around 60–70 mV/decade is assumed over the cathode potential range relevant to PEFC operation. However, it has been suggested by experiments that this approach has to be modified to account for the potential-dependent oxide coverage [183–186]. To implement this coverage, a term is added to the ORR kinetic Eq. (109) [187],

$$i_{\text{ORR}} = -i_{0_{\text{ORR}}} (1 - \Theta_{\text{PtO}}) \left(\frac{p_{O_2}}{p_{O_2}^{\text{ref}}}\right)^{m_0} \left(\frac{a_{\text{HM}}}{a_{\text{HM}}^{\text{ref}}}\right)^4 \exp\left(\frac{-\alpha_c F}{RT}(\eta_{\text{ORR}})\right) \tag{110}$$

where Θ_{PtO} is the coverage of Pt oxide; it should be noted that several oxides can exist and here PtO is taken as an example. There are various methods to calculate PtO using kinetic equations. For example, Yoon and Weber adopt [188]

$$\theta_{PtO} = \frac{\exp\left[\frac{\alpha_a'F}{RT}\eta_{PtO}\right]}{\exp\left[\frac{\alpha_a'F}{RT}\eta_{PtO}\right] + \exp\left[\frac{-\alpha_c'F}{RT}\eta_{PtO}\right]} \tag{111}$$

where

$$\eta_{PtO} = \Phi_1 - \Phi_2 - U_{PtO} \tag{112}$$

and η_{PtO} and U_{PtO} are the Pt-oxide overpotential and equilibrium potential, respectively.

In addition to the above reactions, additional reactions may occur in the catalyst layers including carbon oxidation, Pt dissolution, etc., which are durability concerns and beyond the scope of this chapter [104, 113, 189]. Similarly, one may encounter poisoning of the catalyst sites due to contaminants or even just the ionomer itself [190]. Finally, one should note that the above expressions are for proton-exchange-membrane fuel cells, and for alkaline-exchange-membrane fuel cells the expressions will be different as water is now a reactant and hydroxide species are the mobile charge carriers [191].

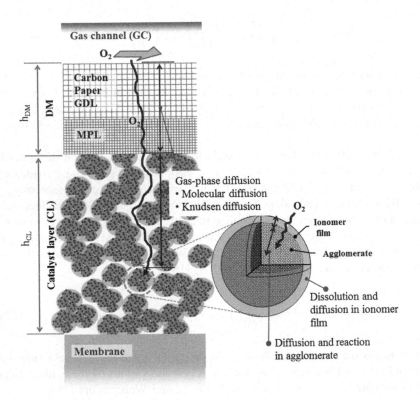

Fig. 12 Schematic of oxygen transport phenomena into and through the cathode catalyst layer (image courtesy of Nobuaki Nonoyama)

5.2 Transport Phenomena

In the catalyst layers, all of the species and phases discussed in previous sections exist in addition to source terms for heat, reactant consumption, product generation, etc. Thus, the conservation and transport equations described in Sects. 3 and 4 for the membrane and porous media, respectively, are used for modeling. Similarly, phenomena such as multiphase flow (discussed in Sect. 4.1) occur, although with the additional inclusion of the reaction source terms. Figure 12 shows the schematic representation of oxygen transport into and through the catalyst layer. Here one can see that multiple diffusion mechanisms occur as well as transport into the reactive agglomerate that may be covered with a film of ionomer or water.

The kinetic expressions, Eqs. 107 and 110 for the HOR and ORR, respectively, result in a transfer current between the electronic-conducting (1) and ionic-conducting (2) phases (see Eq. 10), which is related to the current density in the two phases through Eq. 20 (neglecting double-layer charging)

$$\nabla \cdot \mathbf{i}_2 = -\nabla \cdot \mathbf{i}_1 = a_{1,2} i_{h,1-2} \tag{113}$$

Assuming the ORR is the only reaction that is occurring in the cathode (i.e., crossover and degradation reactions are ignored), the conservation Eq. (10) for oxygen in the cathode can be is written as

$$\nabla \cdot \mathbf{N}_{O_2,G} = -\frac{1}{4F} a_{1,2} i_{0_{ORR}} (1 - \Theta_{PtO}) \left(\frac{p_{O_2}}{p_{O_2}^{ref}}\right)^{m_0} \exp\left(-\frac{\alpha_c F}{RT}(\eta_{ORR})\right) = \frac{1}{4F} \nabla \cdot \mathbf{i}_1 \tag{114}$$

The interfacial area of the catalyst with respect to electrolyte and gaseous reactants, $a_{1,2}$, is often determined by

$$a_{1,2} = \frac{m_{Pt} A_{Pt}}{L} \tag{115}$$

where L is the thickness of the catalyst layer and m_{Pt} and A_{Pt} are the catalyst loading and surface area (which is typically derived from experiment).

The pore structure of catalyst can be classified into intra agglomerate and inter agglomerate pores where the agglomerate is essentially the mesoscale structures that contain the reaction sites (see Fig. 12). The intra-agglomerate pores determine the utilization of Pt catalyst and inter agglomerate pores facilitate the transport of gases and liquid. The distribution of these pores determines the balance between the electrochemical activity with mass-transport phenomena. Describing the structure is the most complex effort to be undertaken in multiscale modeling and is a key future area of integration for different models and modeling scales. Even in the continuum, macroscopic models there are essentially two major length scales that are considered depending on the inter and intra particle/agglomerate interactions. The former or layer length scale is treated using the approaches described in the preceding sections, with the correct effective transport and material properties, including perhaps from mesoscale simulations [170, 171]. Thus, to model the catalyst layer does not require new equations per se. However, for macroscopic simulations, one would like to use the microscopic phenomena but applied at the macroscopic scale. Such an effort can be achieved by using scaling expressions determined through mesoscale and other modeling approaches as mentioned, or by modifying the transfer-current source term to account for diffusional losses at the agglomerate or reaction scale.

5.2.1 Agglomerate Length Scale and Ionomer Films

In this structure, the reactant or product diffuses through possible external films (e.g., ionomer or water) surrounding the agglomerate particle and into the agglomerate itself where it simultaneously reacts and diffuses (see Fig. 12).

To understand the impact of the agglomerate, an effectiveness factor can be used, which is defined as the ratio of the actual reaction rate to the rate if the entire interior surface is exposed to the conditions outside of the particle [18, 192]

$$E = \frac{4\pi r_{agg}^2 \left(-D_{O_2}^{eff}\frac{dc_{O_2}}{dr}\Big|_{r=r_{agg}}\right)}{\frac{4}{3}\pi r_{agg}^3 \left(-k_{s,m_0} c_{O_2,s}^{m_0}\right)} \tag{116}$$

where k_{s,m_0} is the reaction rate of the ORR at the surface conditions,

$$k_{s,m_0} = \frac{a_{1,2} i_{0_{ORR}}}{4F c_{O_2}^{ref m_0}} (1 - \Theta_{PtO}) \exp\left(-\frac{\alpha_c F}{RT}(\eta_{ORR,1-2})\right) \tag{117}$$

Thus, one can write the transfer current as

$$\nabla \cdot \mathbf{i}_2 = a_{1,2} i_{h,1-2} E \tag{118}$$

For a first-order reaction ($m_0 = 1$), the simultaneous diffusion and reaction into a spherical agglomerate (Eqs. 10 and 14) can be solved analytically to yield an effectiveness factor expression of

$$E = \frac{1}{\phi^2}(\phi \coth(\phi) - 1) \tag{119}$$

where ϕ is the dimensionless Thiele modulus

$$\phi = r_{agg} \sqrt{\frac{k_{s,m_0} C_{O_2,s}^{m_0-1}}{D_{O_2}^{eff}}} \tag{120}$$

which is a measure of the reaction rate to the diffusion rate. Similar expressions hold for other reaction orders with variable error [188, 192].

The above does not account for the existence and influence of external films covering the agglomerate. It is known that the ionomer in the catalyst layer forms a thin film that surrounds the active particles [193–196], which is thought to limit performance at low loadings [125, 188]. When accounting for diffusion through the film, one can use the permeation expression (Eq. 73) from the outside gas phase to the surface of the agglomerate

$$N_{O_2} = -D_{film} \nabla c_{O_2} = -\frac{D_{film} H}{RT} \nabla p_{O_2} = -\psi_{O_2,film} \nabla p_{O_2} \tag{121}$$

where H is Henry's constant and the film can be ionomer, liquid water, or perhaps both. Due to the thinness of the films, one can assume a linear flux, thus Eq. 121 can be written as

$$N_{O_2} = \frac{p_{O_2,ext} - p_{O_2,surf}}{R_{O_2,film}} \tag{122}$$

where $R_{O_2,film}$ is the transport resistance of oxygen through ionomer film,

$$R_{O_2,film} = \frac{\delta_{film}}{\psi_{O_2}} \tag{123}$$

and is used since the film thickness and its transport properties are both unknown. At steady state, the flux given by Eq. 121 is equal to the flux due to reaction and diffusion in the agglomerate; therefore, the unknown concentrations can be replaced. Using the resultant expression in the conservation Eq. (20) yields

$$\nabla \cdot i_1 = 4F p_{O_2,ext}^{m_0} \left(\frac{1}{\frac{1}{R_{O_2,film}} + \frac{1}{k_{s,m_0} E}} \right) \tag{124}$$

and a similar one can be derived for the HOR.

The above expression is the governing equation for the transfer current density and includes both the film and agglomerate resistances. Upon inspection, one can see that as the current density increases (i.e., k_{s,m_0} increases), the film resistance becomes more significant and limiting. It should be noted that the transport resistance through the thin film is expected to be higher than the respective resistance of transport through the bulk polymer. The reason is that confinement effects and polymer morphology will change as the thickness of the film decreases down to the tens of nanometers or below [58, 197–200]. While such confinement effects have been observed, the fundamentals of such interactions including their nature, propagation, and impact on various gas and conduction properties are not well understood and are a focus for future study.

Finally, if there is liquid water in the catalyst layer, this is expected to block the reaction sites. While the multiphase equations discussed above account for this effect in terms of transport parameters, they do not in terms of reaction-site blockage. This latter effect can be addressed by having a growing film of water over the agglomerate and using the approach above (i.e. Eq. 124).

5.3   Electrochemical Impedance Spectroscopy

Electrochemical impedance spectroscopy (EIS) is used to understand complex phenomena occurring in a PEFC cell. This technique is closely associated with catalyst layer phenomena since it allows exploration of dominant resistances that occur typically in the catalyst layer and requires electrodes to get a response. While all the previous sections in this chapter discuss about modeling different regions and phenomena, it is worthwhile to have a section on modeling this specific experimental technique. While other techniques are readily modeled with the various modeling equations or derivatives thereof, EIS falls into a special category due to its pervasiveness, ease, and mystery in understanding and interpreting its resulting data, which are compounded by the complexity of the overlapping phenomena. Before looking at how to employ continuum modeling of different phenomena to understand EIS data, let us briefly look at the technique in particular.

Impedance spectroscopy is a perturbation technique used effectively in understanding overlapping phenomena that affect the electrical properties of a system. The idea is that by applying only a small perturbation during operation, the system response can be studied in situ and in a noninvasive way. It can decouple overlapping processes with different rate coefficients. A sinusoidal perturbation is applied to a system operating under steady state and its response is monitored. Depending on the rate of the processes involved in the system the perturbed variable is transmitted across the system. The time lag between the impulse and the response in time domain at different frequencies is proportional to the conductivity of the limiting process within the system. In terms of operation, EIS works by perturbing the current or voltage and watching the response in the other. So, for a voltage oscillation of

$$E_t = E_0 \sin(\omega t) \tag{125}$$

the current response is phase shifted by ϕ to be

Fig. 13 Example of Nyquist
plot and EIS results

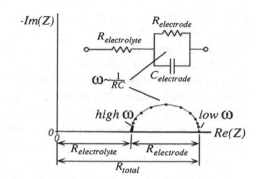

$$I_t = I_0 \sin(\omega t + \phi) \tag{126}$$

and the frequency-dependent EIS is given in terms of real and imaginary parts

$$Z(\omega) = \frac{E}{I} = Z_0(\cos \phi + j \sin \phi) \tag{127}$$

As expected, much of the EIS modeling focuses on the cathode catalyst. When the cathode is studied, usually the whole cell impedance is monitored assuming negligible anodic resistance due to fast HOR kinetics and fixed membrane resistance under fully humidified conditions [201, 202]. Application of EIS to a PEFC is quite important because the overlap of multitudes of phenomena results in a quite difficult decoupling, especially if just relying on only time domain information. Frequency domain information helps in resolving the competing phenomena in a better fashion. However, in PEFCs, there are very few publications that use realistic physical models to describe experimental EIS data, while a majority of EIS works take the easier and relatively less useful route of mapping the data to an equivalent circuit. EIS methods have been employed from characterizing individual components of PEFCs like membrane [21, 203] or catalyst layer [204], to diagnosing the whole stack for operational anomalies. While characterizing the components is relatively easy using simpler physical models, when it comes to understanding the full operation or a stack, it is near impossible to deconvolute the contribution of individual subprocesses due to the overlapping physico-chemical processes with similar time constants.

In general, an EIS spectrum is presented as Nyquist plots as shown in Fig. 13. A Nyquist plot has one or more capacitive loops, and in certain special cases a low-frequency inductive loop is also observed. In the frequency range of $1 \sim 10$ kHz, the EIS spectrum intersects with the real axis called the high-frequency resistance, which indicates the resistance of the cell that includes the ionic membrane resistance, electrical resistances (both bulk and contact), and often the anodic hydrogen oxidation kinetic resistance. From $1 \sim 1000$ Hz range, one or more capacitive loops are observed. The general explanation is that these loops are attributed to oxygen reduction reaction at the cathode. At low overpotentials and fully humidified operating conditions the spectrum typically contains a single loop that is attributed to the oxygen reduction reaction. This loop typically has 45° linear segment at higher frequencies that is attributed to proton transport resistance [3, 205] and oxygen diffusion resistance [206, 207] at the cathode. At higher overpotentials multiple loops are formed. Under certain cases, low-frequency inductance loops were also observed at frequency range below 1 Hz and are attributed to oxygen kinetics involving intermediates, Pt dissolution, and slow water transport across the membrane under low-humidity conditions [201, 208]. Wagner et al. [209] attributed the low-frequency induction loop to surface relaxation process occurring due to CO poisoning at the anode under constant-current conditions.

The simplest method of modeling the EIS spectra by a PEFC is to fit the EIS spectra by an equivalent circuit of electrical elements like resistors, capacitors, constant phase elements, and inductors (for example, see Fig. 11 or 12) [210]. The simplicity offered by equivalent-circuit modeling is also the limitation when it comes to understanding the underlying phenomena in an electrochemical cell. The equivalent circuit does not account for geometry variations, frequency-dependent parameters and variables, physics-based phenomena, etc. To account for these effects rigorously requires a true physics-based continuum model.

To model the impedance, each dependent variable can be written as the sum of its steady-state component $(-)$ and time-varying component $(\sim)$ that is frequency dependent. For example, the current is written as

$$I = \bar{I} + \tilde{I}_t \tag{128}$$

The time-dependent component for a sinusoidal perturbation with fixed amplitude is

$$\tilde{I}_t = I_0 \sin(\omega t) \tag{129}$$

Hence the net current is,

$$I = \bar{I} + \mathrm{Re}\{\tilde{I}e^{j\omega t}\} \tag{130}$$

Similarly, for any variable in the governing equations described in the previous sections, we write

$$x_i = \bar{x}_i + \mathrm{Re}\{\tilde{x}_i e^{j\omega t}\} \tag{131}$$

To determine the frequency-dependent part of each of the variable, each of the variables is expanded in a Taylor series. Assuming the perturbation is small enough that the system responds linearly, the higher order terms are neglected,

$$f(x) = f(\bar{x} + \mathrm{Re}[\tilde{x}e^{jwt}]) = f(\bar{x}) + \frac{df}{dx}\bigg|_{\bar{x}} \mathrm{Re}[\tilde{x}e^{jwt}] \tag{132}$$

Rewriting the above equations in matrix form to evaluate the unknown frequency-dependent terms,

$$
\begin{vmatrix} J_{\text{real}} & -J_{\text{Im}} \\ J_{\text{Im}} & J_{\text{real}} \end{vmatrix} \cdot \begin{vmatrix} \tilde{x}_{\text{real}} \\ \tilde{x}_{\text{Im}} \end{vmatrix} = \begin{vmatrix} G_{\text{real}} \\ G_{\text{Im}} \end{vmatrix}
\tag{133}
$$

Thus, there is only the need to evaluate the time derivative and one can make use of the existing Jacobians to speed up the solution process. From the resulting values, the frequency-dependent EIS is written as the ratio of the frequency-dependent potential and current,

$$
Z = \frac{\tilde{V}}{\tilde{i}}
\tag{134}
$$

By using the governing equations and expressions, one can understand how each variable or property affects the EIS, or one can actually design an algorithm to fit the EIS by changing physical properties and parameters. In terms of computation, the transformations can be done numerically or analytically if possible, and essentially the number of unknowns doubles since each variable now has both a real and an imaginary component. Overall, EIS is a very powerful experimental tool, especially for characterization and trends, but its results are only as meaningful as the model used for its analysis.

6 Summary and Future Outlook

Modeling of polymer-electrolyte fuel cells has come a long way over the last couple of decades, where the simultaneous advent of increased computational power and diagnostic techniques have allowed much greater predictability, complexity, and usefulness of the models. Today, design and optimization strategies can be computationally explored with high fidelity and confidence. A proper understanding also helps in technology development to improve the system in terms of effectiveness and efficiency. Hence, modeling played and will play an important role in the development of PEFCs to be efficient and economic. The above sections in this chapter are meant to provide the background of the underlying physics and their mathematical description such that one can understand the macroscopic phenomena.

However, as discussed throughout, there is still a need to link the macroscopic observables and equations with those at smaller lengthscales in order to provide a truly representative simulation. While such linkages are typically accomplished by transfer of properties, adaptive mechanisms and bidirectional coupling and multi-scaling provide still new opportunities and challenges for modeling PEFCs. Currently continuum models use a number of fitting parameters to correlate properties dependence on certain experimental measurable values. For example, for the membrane, the interaction parameter between polymer and water is extracted from fitting a curve that links the water activity and uptake. Atomistic models can provide a theoretical basis on why the value of the fitted interaction parameter is what it is, and also how to calculate it for new materials. Similarly, the continuum mathematical models need parameters from experimental works for determining transport of charge, energy, and species. Some of these parameters are generally assumed and are not measured because of limitations of suitable methods. The parameter uncertainties coupled with complex PEFC models require validation.

In addition to the scale-coupling opportunities, other macroscopic modeling challenges remain. For example, one needs to understand the sensitivity of the various input parameters (e.g., transport properties) whether they come from experiment or lower-scale models. By knowing the sensitivity, we can see how that matches with experimental results and also provide direction to future activities and research areas. Similarly, this can help to link the modeling results with real-world results in which the systems are much more dynamic and variables with nonuniform properties and stressors. Once this is accomplished, full optimization can be undergone with the model that accounts for full consideration of the cooperative phenomena and material- and operating-design specifications.

In the fashion, acceleration of components development requires correlation of the measurable properties to the performance effect on fuel cells over its lifetime. This is a place where modeling has a huge role to play, since experiments require substantial times to enact. Also, the use of accelerated stress tests and their applicability to real-world and *in-operando* conditions can be correlated through validated cell models. The issue of lifetime predictions dovetails into that of durability modeling, where there is substantial room for improvement and understanding. While there are some degradation-specific models for phenomena such as catalyst dissolution, a majority of models just alter their transport or related properties (e.g., water-uptake isotherm or capillary pressure—saturation relationship) as a function of time. This method is similar to the use of a fitting parameter with time, and more in-depth modeling and knowledge of degradation phenomena are required to increase the usefulness of macroscopic modeling for durability as it is already for performance.

References

1. Perry RH, Green DW (1997) Perry's chemical engineers' handbook. McGraw-Hill, New York
2. CRC Handbook of Chemistry and Physics, CRC Press, Boca Raton, FL (1979)
3. Eikerling M, Kornyshev AA (1998) J Electroanal Chem 453:89
4. Kulikovsky AA, Divisek J, Kornyshev AA (1999) J Electrochem Soc 146:3981
5. Patankar S (1980) Numerical heat transfer and fluid flow. Hemisphere Publishing Corporation,
6. Amphlett JC, Baumert RM, Mann RF, Peppley BA, Roberge PR, Harris TJ (1995) J Electrochem Soc 142:9

7. Sena DR, Ticianelli EA, Paganin VA, Gonzalez ER (1999) J Electroanal Chem 477:164
8. Parthasarathy A, Srinivasan S, Appleby AJ, Martin CR (1992) J Electrochem Soc 139:2530
9. Parthasarathy A, Srinivasan S, Appleby AJ, Martin CR (1992) J Electrochem Soc 139:2856
10. Mosdale R, Srinivasan S (1995) Electrochim Acta 40:413
11. Beattie PD, Basura VI, Holdcroft S (1999) J Electroanal Chem 468:180
12. Lee SJ, Mukerjee S, McBreen J, Rho YW, Kho YT, Lee TH (1998) Electrochim Acta 43:3693
13. Liebhafsky HA, Cairns EJ, Grubb WT, Niedrach LW (1965) In: Fuel cell systems. American Chemical Society, Washington, D.C., p 116 (1965)
14. Shah AA, Luo KH, Ralph TR, Walsh FC (2011) Electrochim Acta 56:3731
15. Bernardi DM, Verbrugge MW (1992) J Electrochem Soc 139:2477
16. Springer TE, Zawodzinski TA, Gottesfeld S (1991) J Electrochem Soc 138:2334
17. Weber AZ (2008) J Electrochem Soc 155:B521
18. Weber AZ, Newman J (2004) Chem Rev 104:4679
19. Franco AA, Guinard M, Barthe B, Lemaire O (2009) Electrochim Acta 54:5267
20. St-Pierre J (2011) J Power Sources 196:6274
21. Olapade PO, Meyers JP, Mukundan R, Davey JR, Borup RL (2011) J Electrochem Soc 158: B536
22. Mao L, Wang C-Y, Tabuchi Y (2007) J Electrochem Soc 154:B341
23. Meng H, Ruan B (2011) Int J Energy Res 35:2
24. Balliet RJ (2010) Chemical Engineering. University of California, Berkeley, p 170
25. Weber AZ, Newman J (2005) ECS Trans 1(16):61
26. Bird RB, Stewart WE, Lightfoot EN (1960) Transport phenomena. Wiley, New York
27. Newman J, Thomas-Alyea KE (2004) Electrochemical systems. Wiley, New York
28. Pintauro PN, Bennion DN (1984) Ind Eng Chem Fundam 23:230
29. Bird RB, Stewart WE, Lightfoot EN (2002) Transport phenomena. Wiley, New York
30. Bear J (1988) Dynamics of fluids in porous media. Dover Publications Inc, New York
31. Dullien FAL (1992) Porous media: fluid transport and pore structure. Academic Press, New York
32. Lampinen MJ, Fomino M (1993) J Electrochem Soc 140:3537
33. Shibata S, Sumino MP (1985) J Electroanal Chem 193:135
34. Hickner MA (2010) Mater Today 13:34
35. Hickner MA (2012) J Polym Sci B Polym Phys 50:9
36. Kreuer KD, Paddison SJ, Spohr E, Schuster M (2004) Chem Rev 104:4637
37. Roche EJ, Pineri M, Duplessix R, Levelut AM (1981) J Polym Sci B Polym Phys 19:1
38. Roche EJ, Pineri M, Duplessix R (1982) J Polym Sci Pol Phys 20:107
39. Gierke TD, Munn GE, Wilson FC (1981) J Polym Sci Polym Phys 19:1687
40. Hsu WY, Gierke TD (1983) J Membr Sci 13:307
41. Fujimura M, Hashimoto T, Kawai H (1981) Macromolecules 14:1309
42. Fujimura M, Hashimoto T, Kawai H (1982) Macromolecules 15:136
43. Kumar S, Pineri M (1986) J Polym Sci B Polym Phys 24:1767
44. Haubold HG, Vad T, Jungbluth H, Hiller P (2001) Electrochim Acta 46:1559
45. Kim MH, Glinka CJ, Grot SA, Grot WG (2006) Macromolecules 39:4775
46. Schmidt-Rohr K, Chen Q (2008) Nat Mater 7:75
47. Elliott JA, Wu D, Paddison SJ, Moore RB (2011) Soft Matter 7:6820
48. Kreuer KD, Portale G (2013) Adv Funct Mater 23:5390
49. Rubatat L, Gebel G, Diat O (2004) Macromolecules 37:7772
50. Gebel G, Aldebert P, Pineri M (1993) Polymer 34:333
51. McLean RS, Doyle M, Sauer BB (2000) Macromolecules 33:6541
52. Aleksandrova E, Hiesgen R, Friedrich KA, Roduner E (2007) Phys Chem Chem Phys 9:2735
53. Takimoto N, Wu L, Ohira A, Takeoka Y, Rikukawa M (2009) Polymer 50:534
54. Van Nguyen T, Nguyen MV, Lin GY, Rao NX, Xie X, Zhu DM (2006) Electrochem Solid State Lett 9:A88
55. Affoune AM, Yamada A, Umeda M (2004) Langmuir 20:6965

56. James PJ, Elliott JA, McMaster TJ, Newton JM, Elliott AMS, Hanna S, Miles MJ (2000) J Mater Sci 35:5111
57. Bass M, Berman A, Singh A, Konovalov O, Freger V (2010) J Phys Chem B 114:3784
58. Bass M, Berman A, Singh A, Konovalov O, Freger V (2011) Macromolecules 44:2893
59. Ochi S, Kamishima O, Mizusaki J, Kawamura J (2009) Solid State Ionics 180:580
60. Jalani NH, Datta R (2005) J Membr Sci 264:167
61. Choi P, Jalani NH, Datta R (2005) J Electrochem Soc 152:E84
62. Morris DR, Sun XD (1993) J Appl Polym Sci 50:1445
63. Zawodzinski TA, Derouin C, Radzinski S, Sherman RJ, Smith VT, Springer TE, Gottesfeld S (1993) J Electrochem Soc 140:1041
64. Takata H, Mizuno N, Nishikawa M, Fukada S, Yoshitake M (2007) Int J Hydrogen Energy 32:371
65. Yau TC, Cimenti M, Bi XTT, Stumper J (2013) J Power Sources 224:285
66. Kusoglu A, Kienitz BL, Weber AZ (2011) J Electrochem Soc 158:B1504
67. Weber AZ, Newman J (2002) In: Van Zee JW, Fuller TF, Gottesfeld S, Murthy M (eds) *Proton Conducting Membrane Fuel Cells* III. The Electrochemical Society Proceeding Series, Pennington, NJ (2002)
68. Bass M, Freger V (2008) Polymer 49:497
69. Freger V (2002) Polymer 43:71
70. Freger V (2009) J Phys Chem B 113:24
71. Pineri M, Volino F, Escoubes M (1985) J Polym Sci Part B Polym Phys 23:2009
72. Hsu WY, Gierke TD (1982) Macromolecules 15:101
73. Mauritz KA, Rogers CE (1985) Macromolecules 18:483
74. Alberti G, Narducci R (2009) Fuel Cells 9:410
75. Kusoglu A, Santare MH, Karlsson AM (2009) Polymer 50:2481
76. Kusoglu A, Savagatrup S, Clark KT, Weber AZ (2012) Macromolecules 45:7467
77. Kreuer KD (2013) Solid State Ionics 252:93
78. Flory PJ (1953) Principles of polymer chemistry. Cornell University Press, Ithaca, p 672
79. Choi P, Jalani NH, Datta R (2005) J Electrochem Soc 152:E123
80. Choi PH, Datta R (2003) J Electrochem Soc 150:E601
81. Laporta M, Pegoraro M, Zanderighi L (1999) Phys Chem Chem Phys 1:4619
82. Thompson EL, Capehart TW, Fuller TJ, Jorne J (2006) J Electrochem Soc 153:A2351
83. Siu A, Schmeisser J, Holdcroft S (2006) J Phys Chem B 110:6072
84. Saito M, Hayamizu K, Okada T (2005) J Phys Chem B 109:3112
85. Futerko P, Hsing IM (1999) J Electrochem Soc 146:2049
86. Yeo RS (1980) Polymer 21:432
87. Kusoglu A, Hexemer A, Jiang RC, Gittleman CS, Weber AZ (2012) J Membr Sci 421:283
88. Weber AZ, Newman J (2004) AIChE J 50:3215
89. Divisek J, Eikerling M, Mazin V, Schmitz H, Stimming U, Volfkovich YM (1998) J Electrochem Soc 145:2677
90. Eikerling M, Kornyshev AA, Stimming U (1997) J Phys Chem B 101:10807
91. Weber AZ, Newman J (2004) J Electrochem Soc 151:A311
92. Choi P, Jalani NH, Datta R (2005) J Electrochem Soc 152:A1548
93. Eikerling MH, Berg P (2011) Soft Matter 7:5976
94. Prausnitz JM, Lichtenthaler RN, Azevedo EG (1999) Molecular thermodynamics of fluid-phase equilibria, 3rd edn. Prentice-Hall Inc, Upper Saddle River, NJ
95. Kusoglu A, Weber AZ (2012) Polymers for energy storage and delivery: polyelectrolytes for batteries and fuel cells. American Chemical Society, p 175
96. Hwang GS, Parkinson DY, Kusoglu A, MacDowell AA, Weber AZ (2013) ACS Macro Lett 2:288
97. Ye XH, LeVan MD (2003) J Membr Sci 221:147
98. Adachi M, Navessin T, Xie Z, Li FH, Tanaka S, Holdcroft S (2010) J Membr Sci 364:183
99. Majsztrik P, Bocarsly A, Benziger J (2008) J Phys Chem B 112:16280
100. Monroe CW, Romero T, Merida W, Eikerling M (2008) J Membr Sci 324:1

101. Satterfield MB, Benziger JB (2008) J Phys Chem B 112:3693
102. Adachi M, Navessin T, Xie Z, Frisken B, Holdcroft S (2009) J Electrochem Soc 156:B782
103. Weber AZ, Newman J (2007) J Electrochem Soc 154:B405
104. Meyers JP, Darling RM (2006) J Electrochem Soc 153:A1432
105. Burlatsky SF, Atrazhev V, Cipollini N, Condit D, Erikhman N (2006) ECS Trans 1:239
106. Liu W, Zuckerbrod D (2005) J Electrochem Soc 152:A1165
107. Kocha SS, Yang JDL, Yi JS (2006) AIChE J 52:1916
108. Promislow K, St-Pierre J, Wetton B (2011) J Power Sources 196:10050
109. Kusoglu A, Santare MH, Karlsson AM, Cleghorn S, Johnson WB (2008) J Polym Sci B Polym Phys 46:2404
110. Nazarov I, Promislow K (2007) J Electrochem Soc 154:B623
111. Wang L, Prasad AK, Advani SG (2011) J Electrochem Soc 158:B1499
112. Huang XY, Solasi R, Zou Y, Feshler M, Reifsnider K, Condit D, Burlatsky S, Madden T (2006) J Polym Sci B Polym Phys 44:2346
113. Borup R, Meyers J, Pivovar B, Kim YS, Mukundan R, Garland N, Myers D, Wilson M, Garzon F, Wood D, Zelenay P, More K, Stroh K, Zawodzinski T, Boncella J, McGrath JE, Inaba M, Miyatake K, Hori M, Ota K, Ogumi Z, Miyata S, Nishikata A, Siroma Z, Uchimoto Y, Yasuda K, Kimijima KI, Iwashita N (2007) Chem Rev 107:3904
114. Lim C, Ghassemzadeh L, Van Hove F, Lauritzen M, Kolodziej J, Wang GG, Holdcroft S, Kjeang E (2014) J Power Sour (2014)
115. Zhou C, Guerra MA, Qiu ZM, Zawodzinski TA, Schiraldi DA (2007) Macromolecules 40:8695
116. La Conti AB, Hamdan M, McDonald RC (2003) In: Vielstich W, Lamm A, Gasteiger HA (eds) Handbook of Fuel Cells: Fundamentals, Technology, and Applications, vol 1. Wiley, New York, p 647 (2003)
117. Kumar M, Paddison SJ (2012) J Mater Res 27:1982
118. Bontha JR, Pintauro PN (1994) Chem Eng Sci 49:3835
119. Booth FJ (1951) J Chem Phys 19:391
120. Pintauro PN, Verbrugge MW (1989) J Membr Sci 44:197
121. Kreuer KD (2002) ChemPhysChem 3:771
122. Kreuer KD (2000) Solid State Ionics 136:149
123. Guzman-Garcia AG, Pintauro PN, Verbrugge MW, Hill RF (1990) AIChE J 36:1061
124. Zamel N, Li XG (2013) Prog Energy Combust Sci 39:111
125. Nonoyama N, Okazaki S, Weber AZ, Ikogi Y, Yoshida T (2011) J Electrochem Soc 158: B416
126. Passalacqua E, Squadrito G, Lufrano F, Patti A, Giorgi L (2001) J Appl Electrochem 31:449
127. Tucker MC, Odgaard M, Yde-Anderson S, Thomas JO (2003) in 203rd Meeting of the Electrochemical Society, Paris
128. Jordan LR, Shukla AK, Behrsing T, Avery NR, Muddle BC, Forsyth M (2000) J Power Sources 86:250
129. Jordan LR, Shukla AK, Behrsing T, Avery NR, Muddle BC, Forsyth M (2000) J Appl Electrochem 30:641
130. Kong CS, Kim D-Y, Lee H-K, Shul Y-G, Lee T-H (2002) J Power Sources 108:185
131. Weber AZ, Newman J (2004) International Communications on Heat and Mass Transfer (in press)
132. Mason EA, Malinauskas AP (1983) Gas transport in porous media: the dusty-gas model. Elsevier, Amsterdam
133. Rosen T, Eller J, Kang J, Prasianakis NI, Mantzaras J, Buchi FN (2012) J Electrochem Soc 159:F536
134. Hwang GS, Weber AZ (2012) J Electrochem Soc 159:F683
135. Martinez MJ, Shimpalee S, Van Zee JW (2009) J Electrochem Soc 156:B80
136. Nam JH, Kaviany M (2003) Int J Heat Mass Transf 46:4595
137. Tomadakis MM, Sotirchos SV (1993) AIChE J 39:397
138. Hwang JJ (2006) J Electrochem Soc 153:A216

139. Burheim OS, Ellila G, Fairweather JD, Labouriau A, Kjelstrup S, Pharoah JG (2013) J Power Sources 221:356
140. Burheim OS, Pharoah JG, Lampert H, Vie PJS, Kjelstrup S (2001) J Fuel Cell Sci Technol 8
141. Morris DRP, Gostick JT (2012) Electrochim Acta 85:665
142. Smith WO (1933) Physics 4:425
143. Leverett MC (1941) Petrol Div Trans Am Inst Mining Metall Eng 142:152
144. Gostick JT, Ioannidis MA, Fowler MW, Pritzker MD (2009) J Power Sources 194:433
145. Gostick JT, Fowler MW, Ioannidis MA, Pritzker MD, Volfkovich YM, Sakars A (2006) J Power Sources 156:375
146. Weber AZ (2010) J Power Sources 195:5292
147. Gostick JT (2011) ECS Trans 41:125
148. Medici EF, Allen JS (2011) ECS Trans 41:165
149. Medici EF, Allen JS (2010) J Electrochem Soc 157:B1505
150. Shi Y, Xiao JS, Pan M, Yuan RZ (2006) J Power Sources 160:277
151. Weber AZ, Newman J (2006) J Electrochem Soc 153:A2205
152. Weber AZ, Hickner MA (2008) Electrochim Acta 53:7668
153. Kim S, Mench MM (2009) J Electrochem Soc 156:B353
154. Burheim O, Vie PJS, Pharoah JG, Kjelstrup S (2010) J Power Sources 195:249
155. Khandelwal M, Mench MM (2006) J Power Sources 161:1106
156. Mukherjee PP, Kang QJ, Wang CY (2011) Energy Environ Sci 4:346
157. Gao Y, Zhang XX, Rama P, Liu Y, Chen R, Ostadi H, Jiang K (2012) Fuel Cells 12:365
158. Kim SH, Pitsch H (2009) J Electrochem Soc 156:B673
159. Hao L, Cheng P (2009) J Power Sources 186:104
160. Wang LP, Afsharpoya B (2006) Math Comput Simul 72:242
161. Mantzaras J, Freunberger SA, Buchi FN, Roos M, Brandstatter W, Prestat M, Gauckler LJ, Andreaus B, Hajbolouri F, Senn SM, Poulikakos D, Chaniotis AK, Larrain D, Autissier N, Marechal F (2004) Chimia 58:857
162. Das PK, Grippin A, Kwong A, Weber AZ (2012) J Electrochem Soc 159:B489
163. Andreaus B, Eikerling M (2009) Device Mater Model Pem Fuel Cells 113:41
164. Kienitz B, Yamada H, Nonoyama N, weber AZ (2011) J Fuel Cell Sci Technol 8:011013
165. Zhao Z, Castanheira L, Dubau L, Berthome G, Crisci A, Maillard F (2013) J Power Sources 230:236
166. Arisetty S, Wang X, Ahluwalia RK, Mukundan R, Borup R, Davey J, Langlois D, Gambini F, Polevaya O, Blanchet S (2012) J Electrochem Soc 159:B455
167. Swider-Lyons KE, Campbell SA (2013) J Phys Chem Lett 4:393
168. Franco AA, Tembely M (2007) J Electrochem Soc 154:B712
169. Malek K, Franco AA (2011) J Phys Chem B 115:8088
170. Patel A, Artyushkova K, Atanassov P, Colbow V, Dutta M, Harvey D, Wessel S (2012) J Vac Sci Technol A, 30
171. Harvey D, Pharoah JG, Karan K (2008) J Power Sources 179:209
172. Zenyuk IV, Litster S (2012) J Phys Chem C 116:9862
173. Chan K, Eikerling M (2011) J Electrochem Soc 158:B18
174. Vetter KJ (1967) Electrochemical kinetics. Academic Press Inc, New York
175. Wang JX, Springer TE, Adzic RR (2006) J Electrochem Soc 153:A1732
176. Neyerlin KC, Gu WB, Jorne J, Gasteiger HA (2007) J Electrochem Soc 154:B631
177. Neyerlin KC, Gu WB, Jorne J, Gasteiger HA (2006) J Electrochem Soc 153:A1955
178. Appleby AJ (1970) J Electrochem Soc 117:328
179. Kinoshita K (1992) Electrochemical oxygen technology. Wiley, New York
180. Parthasarathy A, Dave B, Srinivasan S, Appleby AJ, Martin CR (1992) J Electrochem Soc 139:1634
181. Parthasarathy A, Srinivasan S, Appleby AJ, Martin CR (1992) J Electroanal Chem 339:101
182. Uribe FA, Springer TE, Gottesfeld S (1992) J Electrochem Soc 139:765
183. Wang JX, Zhang J, Adzic RR (2007) J Phys Chem A 111:12702
184. Liu Y, Mathias M, Zhang J (2010) Electrochem. Solid-State Lett 12:B1

185. Caremans TP, Loppinet B, Follens LRA, van Erp TS, Vermant J, Goderis B, Kirschhock CEA, Martens JA, Aerts A (2010) Chem Mater 22:3619
186. Gottesfeld S (2008) ECS Trans 6:51
187. Stamenkovic VR, Mun BS, Arenz M, Mayrhofer KJJ, Lucas CA, Wang G, Ross PN, Markovic NM (2007) Nat Mater 6:241
188. Yoon W, Weber AZ (2011) J Electrochem Soc 158:B1007
189. Darling RM, Meyers JP (2005) J Electrochem Soc 152:A242
190. Subbaraman R, Strmcnik D, Stamenkovic V, Markovic NM (2010) J Phys Chem C 114:8414
191. Zhang H, Ohashi H, Tamaki T, Yamaguchi T (2012) J Phys Chem C 116:7650
192. Fogler HS (1992) Elements of chemical reaction engineering. Prentice-Hall, Upper Saddle River, NJ
193. Xie J, Wood DL, More KL, Atanassov P, Borup RL (2005) J Electrochem Soc 152:A1011
194. Mathias MF, Makharia R, Gasteiger HA, Conley JJ, Fuller TJ, Gittleman CJ, Kocha SS, Miller DP, Mittelsteadt CK, Xie T, Van SG, Yu PT (2005) Electrochem Soc Interface 14:24
195. Mashio T, Malek K, Eikerling M, Ohma A, Kanesaka H, Shinohara K (2010) J Phys Chem C 114:13739
196. Xie Z, Navessin T, Shi K, Chow R, Wang QP, Song DT, Andreaus B, Eikerling M, Liu ZS, Holdcroft S (2005) J Electrochem Soc 152:A1171
197. Modestino MA, Kusoglu A, Hexemer A, Weber AZ, Segalman RA (2012) Macromolecules 45:4681
198. Eastman SA, Kim S, Page KA, Rowe BW, Kang SH, DeCaluwe SC, Dura JA, Soles CL, Yager KG (2012) Macromolecules 45:7920
199. Dishari SK, Hickner MA (2012) ACS Macro Lett 1:291
200. Modestino MA, Paul DK, Dishari S, Petrina SA, Allen FI, Hickner MA, Karan K, Segalman RA, Weber AZ (2013) Macromolecules 46:867
201. Wiezell K, Holmstrom N, Lindbergh G (2012) J Electrochem Soc 159:F379
202. Springer TE, Zawodzinski TA, Wilson MS, Gottesfeld S (1996) J Electrochem Soc 143:587
203. Debenjak A, Gasperin M, Pregelj B, Atanasijevic-Kunc M, Petrovcic J, Jovan V (2013) Strojniski Vestnik-J Mech Eng 59:56
204. Eikerling M, Kornyshev AA (1999) J Electroanal Chem 475:107
205. Raistrick ID (1990) Electrochim Acta 35:1579
206. Jaouen F, Lindbergh G (2003) J Electrochem Soc 150:A1699
207. Kulikovsky AA (2012) J Electroanal Chem 669:28
208. Schneider IA, Bayer MH, von Dahlen S (2011) J Electrochem Soc 158:B343
209. Wagner N, Schulze M (2003) Electrochim Acta 48:3899
210. Orazem ME, Tribollet B (2008) Electrochemical impedance spectroscopy. Wiley, Hoboken, NJ

Mathematical Modeling of Aging of Li-Ion Batteries

Charles Delacourt and Mohammadhosein Safari

Abstract The recent interest in full and hybrid electric vehicles powered with Li-ion batteries has prompted for in-depth battery aging characterization and prediction. This topic has become popular both in academia and industry battery research communities. Because it is an interdisciplinary topic, different methods for aging studies are being pursued, ranging from black box types of approaches from the electrical engineering community all the way to physics-based methods mainly brought about by the chemical engineering community. This chapter describes an overall methodology for aging characterization and prediction in Li-ion batteries based on physics-based modeling. In a first section, the typical aging phenomena in LIBs are reviewed along with their effects on the cell internal balancing and performance loss. In a second section, the physics-based models used for aging studies are presented, which includes both the performance models (i.e., aging-free) and aging models. In a third section, the typical aging experiments and characterization methods are introduced, along with their analysis with the physics-based models. Finally, the last section presents an outlook of physics-based aging modeling.

1 Introduction

During the last few decades, increased concern over the environmental impact of the petroleum-based transportation infrastructure has led to a renewed interest in an electric transportation infrastructure. Electric vehicles (EVs) powered by rechargeable Li-ion batteries (LIBs) substitute for the conventional internal combustion engine automobiles in this environmentally friendly infrastructure. A lithium-ion

C. Delacourt (✉)
Laboratoire de Réactivité et de Chimie des Solides, CNRS UMR 7314,
Université de Picardie Jules Verne, Amiens, France
e-mail: charles.delacourt@u-picardie.fr

M. Safari
Department of Chemistry, University of Waterloo, Waterloo, ON, Canada

© Springer-Verlag London 2016
A.A. Franco et al. (eds.), *Physical Multiscale Modeling and Numerical Simulation of Electrochemical Devices for Energy Conversion and Storage*,
Green Energy and Technology, DOI 10.1007/978-1-4471-5677-2_5

battery provides the portability of stored chemical energy with the ability to deliver this energy as electricity with high conversion efficiency and no gaseous exhaust.

The Li-ion secondary battery was first commercialized by Sony in 1991 [1]. This so-called rocking chair secondary battery is based on the reversible exchange of lithium between two insertion materials that operate at different potentials. The active materials are embedded into porous composite electrodes. An electron-conductive additive (e.g., carbon black) ensures a sufficient electronic conductivity of the electrode and a polymer binder (e.g., vinylidene polyfluoride) is necessary for the electrode components to hold together. The composite mixture is coated onto a current collector which consists of a copper foil at the anode and of an aluminum foil at the cathode. A porous polymer membrane (e.g., polypropylene) is inserted in between the two electrodes. The pores of the electrodes and of the separator are filled with a liquid electrolyte generally made up of a mixture of linear and cyclic carbonates (e.g., diethyl carbonate and ethylene carbonate) and a lithium salt (e.g., lithium hexafluorophosphate). A schematic of the elementary sandwich of the cell is represented in Fig. 1. During discharge, the active material at the anode deinserts lithium, whereas that at the cathode inserts Li. The opposite occurs on charge.

Unlike the majority of other secondary batteries, several LIB chemistries (i.e., based on different electrode active materials) are commercially available. The selection of a battery chemistry is mainly guided by the end-user application, in terms of performance, cost, lifetime, etc. There are essentially two chemistries at the

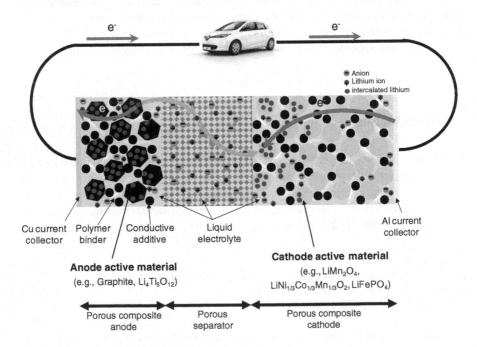

Fig. 1 Schematics of the cell sandwich during discharge

anode, namely graphite and lithium titanate $Li_4Ti_5O_{12}$; Graphite is by far the most popular, while lithium titanate usage is mostly restricted to high-power applications. At the cathode, three classes of materials are used, namely the layered oxides of general formulae $Li[M]O_2$, the spinel oxides of general formulae $Li[M]_2O_4$, and the lithium iron phosphate $LiFePO_4$. In the above formula, M denotes a transition metal element such as Co, Mn, and Ni, as well as some inactive elements like Al, Mg, etc. $LiCoO_2$ (LCO) is the most famous layered material and was used in the first ever commercialized LIB in 1991 [1]. $LiNi_{0.8}Co_{0.15}Al_{0.05}O_2$ (NCA) and $LiNi_{0.33}Mn_{0.33}Co_{0.33}O_2$ (NMC) are other layered materials that have been developed more recently and are also being used in commercial batteries. $LiMn_2O_4$ (LMO) is the most common spinel oxide material and is frequently substituted in order for its stability to be improved [2]. In commercial LIBs, it is common to use mixtures of cathode materials. In particular, LMO is often mixed with a layered material such as NMC or NCA [3, 4]. In addition to a fine tuning of the electrode performance in terms of power and energy density [5], it allows for an improved stability of the LMO, as briefly discussed later on in the chapter. Just like for the negative electrode, the selection of a cathode material depends on the end-user application. The reader is invited to refer to Ref. [6] for a detailed comparison of the active materials and their most common applications in the electric transportation sector.

Long lifespan is a necessary requirement for LIBs that are targeted for transportation and stationary applications. Service life of 10 and 15 years are expected for EVs and hybrid electric vehicles (HEVs), respectively. In terms of cycling, a lifetime up to 3000 deep cycles at room temperature can be requested for high-energy applications (typical for EVs) [7]. Unfortunately, the performance of LIBs declines over time as a consequence of a variety of aging processes. The main objective of modeling battery aging is to predict the fade of the battery performance (i.e., state of health) over a time frame typical of that targeted for the application, which exceeds that of aging experiments done in the lab. The general methodology consists of an in-depth analysis of the aging data that are collected over a limited time frame (typically one to three years), the results of which allow for the derivation of an aging model (Fig. 2). The aging model is then used to extrapolate the cell performance fade to a time frame of interest for the application.

As one expects, accurate battery life prediction is critical to the automotive and stationary sectors, and constitute a necessary input parameter in economic models of an EV/HEV or a stationary storage unit [8]. In its simplest form, the aging model would merely consist of an empirical correlation of the battery capacity and internal resistance as a function of time and a number of aging factors such as current, temperature, state of charge, etc. The state-of-health prediction results from an empirical extrapolation of the correlation over time. A deeper analysis involves a performance model of the battery, in which input parameters are updated to match the battery performance loss over the course of the aging experiments. In this case, the aging model may consist of empirical correlations on each model parameter, in a similar fashion as it is done for cell capacity or internal resistance, just like for the simplest model, as discussed above. The performance model can range from fully

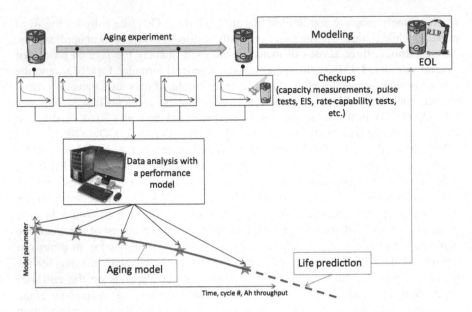

Fig. 2 General methodology for battery aging studies based on experiments and modeling

empirical models (e.g., artificial neural networks) to semi-empirical models (e.g., equivalent electrical circuits) and all the way to physics-based models. The latter category of models relies on the underlying phenomena inside the battery such as mass transport and electrochemical kinetics. A peculiarity of using a physics-based model for aging analysis is the possibility of implementing aging phenomena directly as governing equations into the performance model thus turning it into a physics-based *aging* model [9–13]. This is the most sophisticated approach for aging prediction, because no empirical extrapolation is needed. Besides, it is the most tedious because a comprehensive knowledge of the aging phenomena is required, which calls for in-depth analysis of the cell components. In addition, proper mathematical description of aging phenomena is needed once aging phenomena are unfolded. In this chapter, we restrict ourselves to LIB aging characterization and prediction using physics-based models.

In the first section, the typical aging phenomena in LIBs are detailed along with their effects on the cell internal balancing and performance loss. In the second section, the physics-based models used for aging studies are presented, which includes both the performance models (i.e., aging-free) and aging models. In the third section, the typical aging experiments and characterization methods are introduced, along with their analysis with the physics-based models. Finally, the last section presents an outlook of physics-based aging modeling.

2 Brief Overview of the Degradation Phenomena in Li-Ion Batteries

As any other electrochemical system, a LIB undergoes a progressive decrease of its storage capacity, along with an increase of its impedance, which directly translates into a decrease of its energy and power over time, respectively. There are numerous degradation phenomena, and hence it is hard to evaluate the precise contribution of each of them to the performance fade. Furthermore, it is frequent that some phenomena are coupled with each other, would it be within a single electrode or between the two electrodes (i.e., shuttle-type mechanism), which makes the overall analysis intricate. The objective of this section is not to go through a comprehensive description of all aging phenomena, but instead to present the different types of aging phenomena and how they impact the battery performance. Comprehensive reviews of aging phenomena can be found in Refs. [14–16].

2.1 Aging at the Anode

As mentioned in the introduction, the vast majority of commercially available LIBs contain graphite as anode active material. The potential of Li insertion into graphite is lower than that of the reductive decomposition of the carbonate mixtures typically used as electrolytes. However, a passive film, named the solid electrolyte interphase (SEI) [17], forms at the electrode particle surface, thanks to the accumulation of the decomposition products from electrolyte reduction during the first charge or so of the battery. This SEI is ideally impermeable to both electrons and solvent molecules and therefore nearly suppresses further solvent decomposition. It is, however, permeable to lithium ions, thus allowing for the insertion/deinsertion reaction to proceed. Because of the electrons consumed in the SEI formation, the internal cell balancing is modified, as illustrated in Fig. 3.

Initially, just after cell assembly, the cathode is fully lithiated and the graphitic anode is fully delithiated (red bullet in Fig. 3a). During the first charge of the cell, some electrons are involved in the electrolyte decomposition reaction(s) (i.e., side reaction) rather than for Li insertion into the graphite (i.e., main reaction). It yields a shift of the graphite curve toward the left in Fig. 3b, with that shift being proportional to the charge consumed in the side reaction(s). As indicated in Fig. 3c, the cathode does not get back to a fully lithiated composition at the end of the first discharge, which is the result of the change in cell balancing arising from the side reaction at the graphitic anode.

Although LIBs existence is tied to that of the SEI, it also turns out that the SEI is the primary cause of LIB aging. Side reactions are never totally suppressed, and consequently, the SEI keeps on growing over the battery life, and the associated loss of cyclable lithium leads to capacity fade through a progressive debalancing of the cell, as illustrated in Fig. 4b.

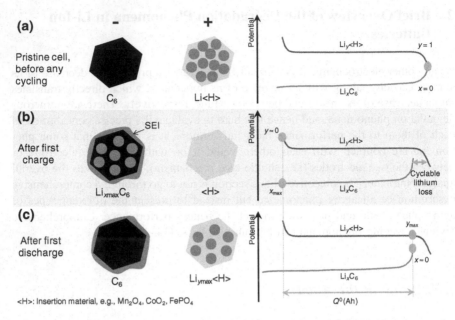

Fig. 3 First charge/discharge cycle of a LIB, showing the effect of the SEI formation on the internal balancing of the cell

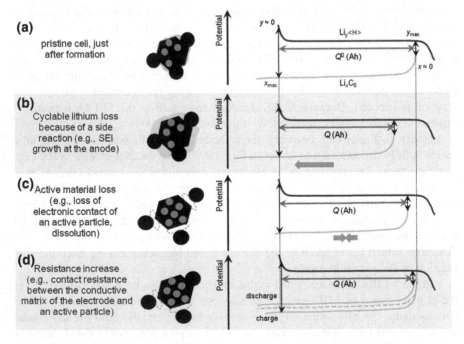

Fig. 4 Effects of different aging phenomena on the cell internal balancing and cell capacity

The SEI growth is promoted under certain operating conditions, as summarized below:

- At elevated storage or cycling temperature, partial dissolution or delamination of the SEI may occur, thus favoring electrolyte decomposition [18]. Elevated temperature also promotes cathode material dissolution, as is especially true for LMO, which releases metallic ions in the electrolyte. Ions diffuse away from the cathode toward the anode and modify the SEI, which then becomes electron conducting and electrochemically active, thereby impairing its passive function toward electrolyte decomposition [19, 20].
- Charging the battery at low temperature and/or overcharging, it can lead to lithium plating at the surface of the graphite particles [21]. Plated lithium either reinserts into the graphite, which is then neutral with regard to aging, or reacts with the electrolyte, in which case it yields to SEI growth and contributes to the cell debalancing. Other conditions, related to the initial cell design, can also lead to lithium plating, such as an excess of cathode material and/or a cathode laminate that extends past the size of the anode laminate [22].
- The repeated cycling of the battery in a large range of SOC can result in crack formation and fractures within the SEI, as a consequence of the volume change of the active material with the lithium content (e.g., about 10 % for graphite between C_6 and LiC_6). Fractures in the SEI account for newly exposed surface of the graphite particles to the electrolyte and therefore promote electrolyte decomposition [23].

In addition to the loss of cyclable lithium, the SEI growth accounts for an increase of the anode impedance, which contributes to an additional capacity loss of the cell when measured under nominal current conditions. This effect is illustrated in Fig. 4d. The SEI impedance might increase at elevated temperature, because SEI is prone to composition changes with the partial conversion of organic components into less Li-ion conductive inorganic components [24]. The main degradation phenomena associated with the SEI and discussed above are summarized in Fig. 5.

Aside of the cyclable lithium loss, the cell may experience a capacity loss as a result of active material particles in either cathode or anode turning inactive (as illustrated for the graphite anode in Fig. 4c). Reasons behind active material loss depend upon the material itself; for instance, solvent co-intercalation in graphite leads to exfoliation of the graphene sheets, hence, damaging the host structure [25]. Graphite particles may also retain their structure while turning inactive because of their ionic or electronic disconnection from the rest of the porous electrode. This is usually referred to as contact loss. Electronic contact losses generally arise from repeated charge/discharge cycling because of the contraction/expansion of the graphite particles that lowers the mechanical integrity of the electrode, up to a point where the electrode delaminates from the current collector.

Lithium titanate, compared with graphite, suffers much less degradation phenomena. First of all, because it operates at a potential of ca. 1.5 V versus Li, solvent reductive decomposition is nearly suppressed, and hence there is almost no cyclable lithium loss. Additionally, the two-phase Li insertion/deinsertion mechanism

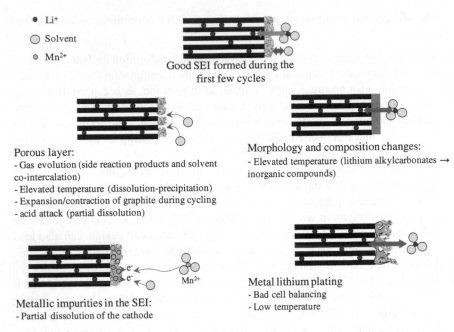

- Li⁺
- Solvent
- Mn²⁺

Good SEI formed during the
first few cycles

Porous layer:
- Gas evolution (side reaction products and solvent co-intercalation)
- Elevated temperature (dissolution-precipitation)
- Expansion/contraction of graphite during cycling
- acid attack (partial dissolution)

Morphology and composition changes:
- Elevated temperature (lithium alkylcarbonates → inorganic compounds)

Metallic impurities in the SEI:
- Partial dissolution of the cathode

Metal lithium plating
- Bad cell balancing
- Low temperature

Fig. 5 Dependence of the SEI properties at the graphite surface on various conditions/phenomena

between $Li_4Ti_5O_{12}$ and $Li_7Ti_5O_{12}$ is nearly strain free, meaning that there is no significant mechanical problem showing up upon cycling [26, 27].

2.2 Aging at the Cathode

Just like for anodes, cathode aging may either originate from the active material itself (e.g., dissolution, phase transitions, structural amorphization), from electrolyte decomposition reactions at the solid/liquid interface (possibly leading to SEI formation), or from degradation at the porous electrode scale (e.g., degradation of the binder and the conductive filler, current collector corrosion, loss of the mechanical integrity) [28]. Degradation phenomena that are the most commonly observed at cathode are summarized in Fig. 6.

Electrolyte oxidative side reactions at the surface of the active material/carbon additive/current collector are frequently observed, in particular at high potential (high state of charge) and/or at elevated temperature. Just like for reductive side reactions at the anode, side reactions at cathode modify the cell balancing, i.e., the cathode curve, represented in blue in Fig. 4, shifts toward the left. Overall, side reactions at cathode have been much less characterized than those at anode. Reaction products may be solid, thus forming an SEI at the surface of the active particles and conductive filler [29]. Gaseous and soluble products have also been

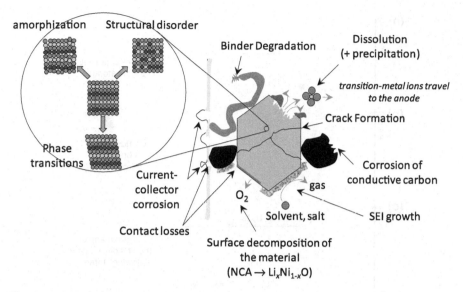

Fig. 6 Typical degradation phenomena at the cathode

reported. Protons are among this latter category and are detrimental to the battery operation because they react with PF_6^- to form HF, which is an impurity known to react with the anode SEI components [30].

The relative rates of side reactions at cathode and anode account for the amounts of reversible and irreversible capacity losses occurring during open-circuit storage of a cell at full charge (Fig. 7), also known as self-discharge. In general, the amount of charge consumed in side reactions at the anode exceeds that at cathode, and as a consequence the reversible capacity fade corresponds to the charge consumed in side reactions at the positive electrode, whereas the irreversible capacity fade is the difference between side reaction charges at anode and at cathode, as illustrated in Fig. 7.

Degradation of cathode materials may result from different processes, most of them being material-dependent. Atomic substitution within the crystal structure, e.g., between lithium and transition metal atoms, is common in some layered oxides, particularly the Ni-rich ones. These defects may impede solid-state Li transport thus increasing the electrode impedance. Some cathode materials undergo irreversible phase transition when driven in a composition range where the initial structure is no longer stable. This is the case of LCO, which exhibits an irreversible transition from the hexagonal toward a monoclinic phase at a composition of about $Li_{0.5}CoO_2$ [31]. In some cases, the phase transition or material decomposition is restricted to the very near surface of the material. NCA undergoes such a surface instability, with the original layered structure transforming to a rocksalt-type $Li_xNi_{1-x}O$ phase, which impedes the insertion/deinsertion reaction.

Dissolution of active material turns out to be a major aging mode at cathode, especially for the spinel-type materials. In LMO, it occurs because of a

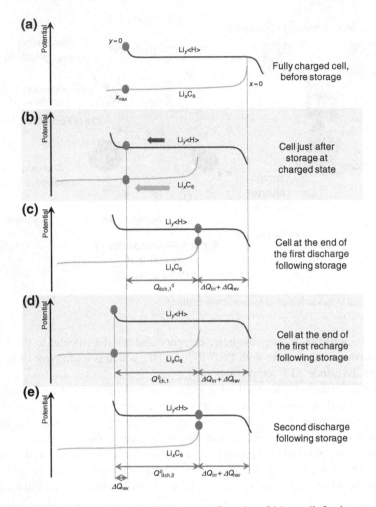

Fig. 7 Schematic representation of self-discharge effects in a Li-ion cell. In the example, the charge involved in reductive side reactions at anode exceeds that involved in oxidative side reactions at cathode

disproportionation reaction of Mn^{3+} into Mn^{4+} (insoluble) and Mn^{2+} (soluble). This reaction is catalyzed by acidic impurities (such as HF) that come from the oxidative side reactions of the electrolyte at high potential [32], as mentioned above. In some cases, dissolved Mn reprecipitates as insulating compounds at the positive electrode (e.g., MnF_2, $MnCO_3$, Mn_xO_y), which yields an increase of the electrode impedance [33]. Dissolved Mn also diffuses to the anode and destabilizes the SEI [19, 20]. The anode contamination with traces of Fe was also reported whenever LFP is used at cathode, but unlike for LMO, it seems that iron dissolution is due to synthesis impurities rather than from LFP particles themselves [34].

3 Mathematical Models

3.1 Performance (Aging-Free) Models

As mentioned in the introduction, building a mathematical model of the cell relying on the underlying physics is the most sophisticated approach for aging analysis and life prediction; unlike with their empirical counterparts, degradation phenomena can be identified and their effects can be quantified. The downside is that an in-depth knowledge of the cell is necessary, both initially and at the different states of health, in order for model parameters to be determined. In most cases, physics-based models are coupled with intrusive characterization techniques that require taking cells apart. Unfortunately, it is commonplace for cell manufacturers to protect their products with a nondisclosure agreement, particularly for cells which are still at a development stage, which prevents the implementation of physics-based models.

3.1.1 Model of the Elementary Sandwich ("Dualfoil")

Dualfoil is the physics-based model that serves as a basis in the battery engineering community [35]. It was developed in the 1990s by the Newman group at University of California Berkeley, USA. It is a mathematical model of the anode/separator/cathode elementary sandwich. The model equations and various model features are described in detail in Ref. [36]. In the present work, only the basic model features are described, for the sake of conciseness. *Dualfoil* is a 1D + 1D (or pseudo 2D) model, i.e., there is a dimension x perpendicular to the elementary sandwich ("macro" domain), and a radial dimension r along the active particles that are assumed spherical and isotropic ("micro" domain). The model is based on the porous electrode theory, meaning that each layer of the sandwich is treated as the superposition of two continua, one representing the electrolyte and another representing the solid matrix. Each continuum is characterized by its volume fraction in the layer and its specific contact area with other continua. Other properties of the continua such as the conductivity are averaged over a volume element which is small compared to the dimension of the layer but large compared to the dimension of the pores [37].

There is an electronic current density $\vec{i}_1$ flowing across the porous electrodes in the solid phase along the x dimension, and an ionic current density $\vec{i}_2$ flows similarly in the liquid phase. The sum of these two current densities is uniform across the electrode, which means that any increase in $\vec{i}_2$ is compensated by a decrease in $\vec{i}_1$. These two current densities are linked through the so-called pore wall flux j_n at the solid/electrolyte interface, according to

$$\nabla \cdot \vec{i_2} = aF j_n \tag{1}$$

with
a Area of the solid/liquid interface per unit volume of the electrode (m^2/m^3),
F Faraday's constant (96,487 C/mol), and
j_n Pore wall flux of lithium [mol/(m^2 s)].

The ionic current density $\vec{i_2}$ is zero at the current collectors of both electrodes, whereas the electronic current density $\vec{i_1}$ is zero at the electrode/separator interface (All current is carried by ions across the separator). The electronic current density $\vec{i_1}$ in the solid matrix of the electrodes is expressed by Ohm's law:

$$\vec{i_1} = -\sigma_1^{\text{eff}} \nabla \Phi_1 \tag{2}$$

with
σ_1^{eff} Effective electronic conductivity of the solid phase of the electrode (S/m) and
Φ_1 Electric potential of the solid phase (V),

whereas the ionic current density $\vec{i_2}$ is expressed using an extended Ohm's law that derives from the concentrated solution theory and takes into account the effect of salt concentration gradient across the liquid phase:

$$\vec{i_2} = -\kappa^{\text{eff}} \nabla \Phi_2 + 2\kappa^{\text{eff}} \frac{RT}{F} \left(1 - t_+^0\right) \left(1 + \frac{d \ln f_\pm}{d \ln c}\right) \nabla \ln c \tag{3}$$

with
κ^{eff} Effective ionic conductivity of the electrolyte (S/m),
Φ_2 Electric potential of the liquid phase (V),
R Ideal gas constant (J/(mol K)),
T Absolute temperature (K),
t_+^0 Transference number of lithium ions in the electrolyte, with respect to the solvent velocity,
$f_\pm$ Mean molar activity coefficient of the electrolyte, and
c Salt concentration in the electrolyte (mol/m^3).

A mass balance on the anion of the salt (e.g., PF_6^-) is used in order to solve for the salt concentration:

$$\frac{\partial(\varepsilon c)}{\partial t} = -\nabla \cdot \vec{N_-} \tag{4}$$

with
t Time (s) and
ε Volume fraction of the liquid phase.

In Eq. (4), $\vec{N}_-$ denotes the flux density of the anion and is replaced by its definition

$$\vec{N}_- \approx -D^{\text{eff}} \nabla c - \left(1 - t_+^0\right) \frac{\vec{i}_2}{F}$$

(5)

with

D^{eff} Effective diffusion coefficient of the salt (m^2/s).

The two boundary conditions of Eq. (4) are obtained by setting $\vec{N}_-$ to zero at both current collectors

Effective parameters in the above equations relate to the volume fraction ε and tortuosity τ of the phase they refer to. For instance, the ionic conductivity of the electrolyte reads

$$\kappa^{\text{eff}} = \frac{\varepsilon}{\tau} \kappa.$$

(6)

It is frequent that the tortuosity is approximated using a function of the volume fraction. Bruggeman equation is the most commonly used relationship and reads $\tau = 1/\sqrt{\varepsilon}$.

The above equations constitute the "macro" model, which is at the sandwich scale. The macromodel connects with the micromodel (at the particle scale) through the pore wall flux j_n, which relates to the rate of the electrochemical reaction (i.e., charge-transfer reaction) according to

$$j_n = \frac{i_n^0}{F} \left\{ \exp\left(\frac{(1-\beta)F}{RT}(\Phi_1 - \Phi_2 - U)\right) - \exp\left(-\frac{\beta F}{RT}(\Phi_1 - \Phi_2 - U)\right) \right\}$$

(7)

with

β Charge-transfer coefficient and

U Equilibrium potential of the electrode, measured versus a Li metal electrode (V).

In Eq. (7), i_n^0 denotes the exchange current density, which is expressed by

$$i_n^0 = F k^0 c^{1-\beta} \left(c_{s,\text{max}} - c_s\right)^{1-\beta} c_s^\beta$$

(8)

with

k^0 Rate constant of the electrochemical reaction $(mol/[m^2 \text{ s } (mol/m^3)^{1.5}])$ and

c_s Li concentration in the active material (mol/m^3), ranging between 0 and $c_{s,\text{max}}$.

The solid-phase lithium concentration at the particle surface is obtained by solving for the Li mass balance in the particle, i.e.,

$$\frac{\partial c_s}{\partial t} = \frac{1}{r^2}\frac{\partial}{\partial r}\left(D_s r^2 \frac{\partial c_s}{\partial r}\right) \tag{9}$$

with
D_s Solid-state Li diffusion coefficient (m²/s).

Two boundary conditions are required when solving for Eq. (9): The lithium flux density is set equal to the pore wall flux j_n at the particle surface and is set to zero at the particle center, by symmetry.

In the Dualfoil model, six independent variables, namely c, c_s, j_n, i_2, Φ_1, and Φ_2, are simultaneously solved for. To this end, the system of ordinary and differential equations summarized in Fig. 8 is used, after differential equations are discretized using the finite-difference method or the control-volume formulation.

3.1.2 Single-Particle Model

The single-particle model is a simplified version of the *Dualfoil* sandwich model. Under low-current operating conditions, limitations across the porous electrodes

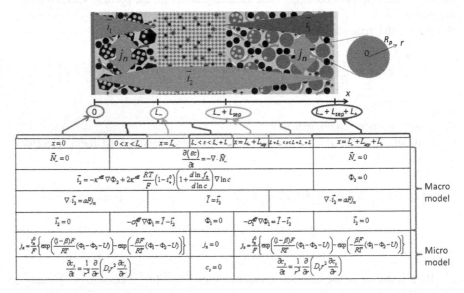

Fig. 8 Elementary sandwich used as the basis for Dualfoil model, which was developed by the Newman research group in the 1990s [35]. Model equations and boundary conditions are provided in the table. Note that boundary conditions for the bottom equation (spherical diffusion in the active particles) are not indicated, for conciseness

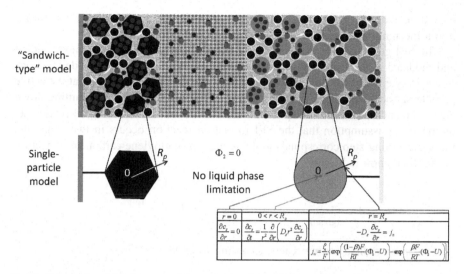

Fig. 9 Schematics of the single-particle approximation, which is well suited to model electrodes operating at low current density (typically <1 C)

(i.e., the electronic conduction in the solid phase and ionic conduction in the liquid phase) become negligible compared with microscale limitations such as charge transfer and solid-state lithium transport. In these conditions, the rate of the electrochemical reaction is identical for all particles that make up the electrodes (providing they have the same properties), and as a consequence, the entire porous electrode can be modeled by a single particle and the set of equations of the "micro" model only (Fig. 9). The single-particle model was originally introduced by Atlung et al. [38]. A review article compares different types of models, bearing different levels of simplification, and including both the sandwich model and the single-particle model [39]. It is demonstrated that the single-particle model is a good approximation of a porous electrode up to rates of about 1 C, with this threshold value depending upon the type of active material making up the electrode. The single-particle model is particularly well suited for aging studies, thanks to its simplicity and its short associated computation time.

3.2 Modeling of Aging Phenomena

In this subsection, we will show how a cell (performance) model can be turned into an aging model by including additional mathematical equations that stand for degradation phenomena. Even though deriving such an aging model seems appealing, it remains a challenge because degradation phenomena are manifold and intricate, and therefore it is difficult to fully characterize them despite the numerous physical and electrochemical techniques available nowadays. Additionally, even once a degradation mechanism is properly unraveled, it is still necessary to turn it into a meaningful mathematical form, which itself is also challenging.

The SEI growth is the degradation phenomenon that has been the most studied and modeled in the battery literature to date. Typically, one or more side reactions of electrolyte reductive decomposition lead to the formation of a layer at the active particle surface. This layer quickly passivates the side reactions by shutting down electronic and electrolyte transport between the particles and the liquid phase. Let us make the assumption that the SEI formation reaction occurs in two steps; the rate-determining step corresponds to the reduction of ethylene carbonate molecules to yield an anion radical.

The anion radical further reacts according to either

or

to yield the ethylene dicarbonate anion, which constitutes the main SEI component, once associated with two lithium ions [13].

Tafel equation is generally used to describe the kinetics of the rate-determining step:

$$i_s = -Fk_{f,s}c_{EC}\exp\left(-\frac{\beta_s F}{RT}(\Phi_1 - R_{SEI}i_t)\right) \qquad (10)$$

with

i_s Current density of the side reaction step (A/m^2),

$k_{f,s}$ Rate constant of the side reaction step (m/s),

c_{EC} Ethylene carbonate (EC) concentration at the particle/SEI layer interface (mol/m^3),

β_s Charge-transfer coefficient of the side reaction step,

R_{SEI} Ionic SEI resistance (Ω m^2), equal to δ/κ_{SEI}, with δ the SEI thickness, and κ_{SEI} the ionic conductivity, and

i_t Total current density (sum of main and side reaction current densities) (A/m^2).

The EC concentration across the SEI film is obtained by solving the following mass balance equation:

$$\frac{\partial c_{EC}}{\partial t} = D_{EC}\frac{\partial^2 c_{EC}}{\partial r^2} - v\frac{\partial c_{EC}}{\partial r} \tag{11}$$

with

D_{EC} EC diffusion coefficient in the SEI (m^2/s) and

v SEI film velocity (m/s).

The SEI velocity directly relates to the current density of the side reaction according to

$$v = \frac{d\delta}{dt} = -\frac{i_s}{n_e F}\frac{M_{SEI}}{\rho_{SEI}} \tag{12}$$

with

n_e Moles of electrons per mole of SEI component formed,

M_{SEI} Molecular weight of the SEI component (kg/mol), and

ρ_{SEI} Density of the SEI component (kg/m^3).

Two boundary conditions are required for Eq. (11) to be solved. The flux density of EC is set equal to the side reaction rate at the particle/SEI interface and the EC concentration is set at the SEI/electrolyte interface assuming that $c_{EC} = \varepsilon_{SEI}c_{EC}^0$, with c_{EC}^0 the EC concentration in the bulk of the electrolyte and ε_{SEI} the volume fraction of electrolyte in the SEI. The SEI growth mechanism that is considered herein is summarized in Fig. 10.

Figure 11 shows a good agreement between the simulated SEI thickness and that derived from experimental data by considering that electrons consumed in the SEI formation and growth correspond to the sum of irreversible and reversible capacity losses (i.e., $\Delta Q_{irr} + \Delta Q_{rev}$ in Fig. 7). The parameter sensitivity study shows that the SEI growth is under mixed control, i.e., EC diffusion across the SEI and side reaction kinetics both limit its growth over time.

Many other models for SEI growth have been published in the literature. In some of them, the SEI growth is assumed to be under diffusion control only (i.e., solvent diffusion across the SEI) [41], whereas in others, it is assumed to be kinetically controlled [12], meaning that no transport limitations are included. Other models rely on a totally different physical representation, wherein the SEI is "leaky" as a result of a non-zero electronic conductivity [11, 42] instead of being sparingly

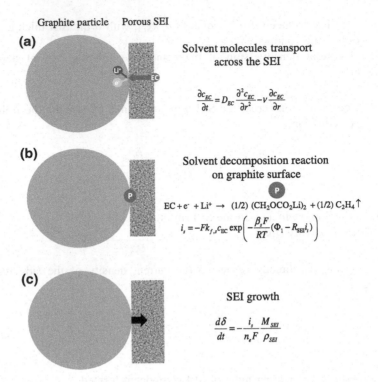

Fig. 10 Physical representation of the SEI formation and growth, and associated equations

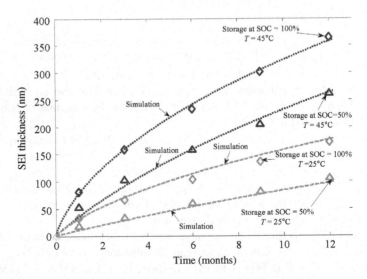

Fig. 11 Evolution of the SEI thickness for a set of graphite/LiFePO₄ cells that are calendar-aged at two different SOCs and two different temperatures [40]. *Markers* Values estimated from experimental data; *Lines* Model simulations

permeable to solvent molecules, like in the above example. Details about these different SEI modeling papers are discussed in the introduction of Ref. [43].

Other degradation phenomena have been turned into mathematical equations and included into physics-based models. For instance, the solvent oxidation at the cathode was treated using Tafel equation, in a similar fashion as it was done for solvent reduction at the anode [9, 43]. In other studies, the lithium metal plating at the anode in overcharge condition or at low temperature was modeled [10, 44]. Because the reaction kinetics is fast for Li plating/stripping, Butler–Volmer equation was used in this case.

Active material loss was the subject of many less modeling studies, in comparison with those focusing on cyclable lithium losses arising from side reactions, as detailed above. The origin of active material losses is generally hard to determine, which makes it more difficult to cast into a mathematical form. For instance, it is well known that the repeated cycling of an active material undergoing volume changes during Li insertion/deinsertion leads to a deterioration of the mechanical properties of the electrode, and to a progressive disconnection of active particles from the conductive matrix, which then become inactive. To some extent, cracks may even form because of diffusion-induced stress in materials undergoing large volume changes during Li insertion/deinsertion. Accordingly, physics-based models of diffusion-induced stress have been developed in an attempt to study the fracture in the intercalation materials [45]. These complex models are helpful to provide design guidelines in order to minimize the mechanical degradation but are not yet comprehensive enough to be put in a direct correlation with the capacity and power loss [46]. To account for the overall active material loss in an aging model, one may start with a simple empirical correlation of the form [43]

$$\frac{\partial \varepsilon}{\partial t} = f(j_n, T, \bar{c}_s), \tag{13}$$

with
ε Volume fraction of the active material and
$\bar{c}_s$ Average lithium concentration in active material (mol/m^3).

The input parameters of this correlation are empirical in nature and can be determined from design of experiments in which the influence of different stress factors (e.g., current, temperature, active particle composition) is probed.

A simple phenomenological treatment of both cyclable lithium and active material loss is possible upon application of evolutionary models of active surface area [47]. The active area increases due to the fracture of particles and decreases because of the particle isolation:

$$a = a_i + a_f - a_{iso} \tag{14}$$

with

a_i Initial active surface area per unit volume of the electrode (m^2/m^3),

a_f Surface area formed upon particle fracture per unit volume of the electrode (m^2/m^3), and

a_{iso} Isolated area due to the fracture per unit volume of the electrode (m^2/m^3).

In analogy to mechanical fatigue of metallic materials, the following equation was used to express the surface area arising from fracture:

$$\frac{d\left(\frac{a_f}{a_i}\right)}{dN} = k_f \tag{15}$$

with

N Cycle number and

k_f Evolution parameter for fracture.

k_f is a material parameter and depends on the strain energy, SOC, ΔSOC, temperature, current, and particle size. The rate of change in surface of isolated area is assumed to be proportional to the total available area

$$\frac{da_{iso}}{dN} = k_{iso}a \tag{16}$$

with

k_{iso} Evolution parameter for isolation.

The following equation results upon combination of Eqs. (15) and (16) for a case where the evolution parameters (i.e., k_f and k_{iso}) are taken to be constant:

$$\frac{a_{iso} - a_f}{a_i} = \left(1 - \frac{k_f}{k_{iso}}\right)\left(1 - e^{-k_{iso}N}\right) \tag{17}$$

In the framework of the above-mentioned evolutionary model, the mechanical degradation (i.e., fracture and isolation) and the electrochemical side reactions (e.g., SEI formation) are linked together. In this case, the side reactions occur on both initial (a_i) and fresh surface exposed through fracture (a_f).

Other phenomena responsible for active material loss, such as dissolution, are more straightforward to model. Let us consider the spinel cathode and assume that dissolution occurs according to the Hunter's reaction:

$$4H^+ + 2LiMn_2O_4 \Leftrightarrow 2Li^+ + Mn^{2+} + 3/2Mn_2O_4 + 2H_2O.$$

The rate of active material loss directly relates to the rate of the above reaction according to [48]

$$\frac{\partial \varepsilon}{\partial t} = -a\bar{V}r_{\text{diss}} \tag{18}$$

with
$\bar{V}$ Molar volume of LMO (m^3/mol) and
r_{diss} Rate of the dissolution reaction [mol/(m^2 s)], expressed with a kinetic expression.

In case some reaction products are soluble in the electrolyte (e.g., Mn^{2+} ions in the above reaction), an additional mass balance is required in the sandwich model for each soluble species. This is of special interest when some soluble species are involved in reactions at both electrodes. Because these species are generally dilute, their flux density is approximated to

$$\vec{N}_i \approx -D_i^{\text{eff}}\nabla c_i \tag{19}$$

with
D_i^{eff} Effective diffusion coefficient of species i in the electrolyte (m^2/s) and
c_i Concentration of species i in the electrolyte (mol/m^3).

The migration term is neglected because the electrolyte contains the highly concentrated lithium salt that acts as a supporting salt. More information about the consideration of soluble species either from side reactions and genuinely present as contaminants in the battery model are available in Refs. [14, 36, 49].

Significant passivation and/or loss of the active material might curtail the effective transport of charge/mass in the electrode. In such instances, empirical equations can be used to update the transport properties over the course of aging. The following equation is an example where the solid-state lithium diffusion coefficient changes as a result of inactive film formation on the surface of the active particles [50]:

$$D_s = D_s^0\left[1 - \left(\frac{\varepsilon^0 - \varepsilon}{\varepsilon^0}\right)^n\right], \tag{20}$$

with
D_s Instantaneous solid-state Li diffusion coefficient (m^2/s),
D_s^0 Initial solid-state Li diffusion coefficient (m^2/s),
ε Instantaneous volume fraction of the active material,
ε^0 Initial volume fraction of the active material, and
n Empirical factor.

4 Model-Aided Analysis of Battery Aging

4.1 Typical Aging Experiments and Characterization

4.1.1 Aging protocols

Typically, aging protocols used in the lab are based on either cycling or storage of the Li-ion cells. Some protocols combine both cycling and storage within a same aging test. Storage tests are popular because in many applications, the battery is left at rest during substantial amounts of time. For instance, it is estimated that an EV spends about 95 % of the time in parking mode. Overall, storage tests are simpler than those of cycling because there are many less operating parameters that can be varied and much less maintenance needed. SOC and temperature are the only two parameters which are generally controlled. There are two types of storage tests; either the cell is set to a state of charge and left at open circuit or it is maintained at a certain potential using a potentiostat (i.e., floating). With the first method, the cell undergoes progressive self-discharge during storage, and as a consequence aging tends to slow down over time. Hence, the cell is brought back to its original state of charge at some time interval (i.e., reSOCing). The number of cell reSOCings during the aging test may influence the results of the test. ReSOCing is generally combined with the cell performance characterizations aimed to evaluate the cell state of health (SOH). With the floating method, there is no reSOCing needed.

Cycling tests are way more diversified and essentially depend upon the application the battery is targeted for. The main operating parameters that are varied when designing cycling tests are the current or power of the cycles/microcycles, the cycling temperature, and the SOC window. The type of cycling profiles ranges from regular constant-current charge/discharge cycles all the way to complex cycles that are specific of the battery usage. Constant-current cycling allows comparing the aging data with others typically reported in the literature for similar or different types of cells/chemistries. Battery manufacturers generally rely on capacity retention during constant-current cycling as a metrics for battery life span.

Complex cycling for EV batteries, such as illustrated in Fig. 12, generally consists of the succession of current or power microcycles that are made up of discharging and charging events. Discharging events surrogate vehicle acceleration whereas charging ones feature regenerative breaking. The cell is generally discharged by repeating the (overall discharging) microcycles and it is recharged under constant current/power, just like a charger would do. Similar types of complex profiles are used for a PHEV or HEV driving profiles, except that corresponding current or power densities lie in a different range, which increases from EV to PHEV and to HEV. Another difference lies in that the operating SOC window decreases from EV to PHEV to HEV. Sometimes, HEV profiles are simply made up of the succession of discharging and charging microcycles swinging the SOC in a narrow range. To the extreme, it is simply made up of the repetition of charge-neutral microcycles around a constant average SOC. Setting the cell temperature during cycling experiments is

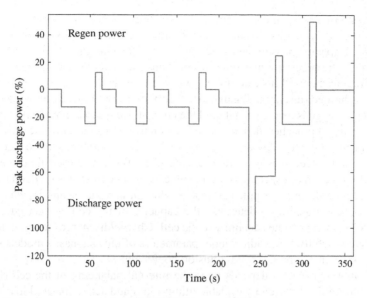

Fig. 12 Dynamic stress test cycle of the USABC manual [51]

trickier than for storage experiments; because of Joule heating, the cell temperature may differ substantially from that of the temperature chamber in which it is located. This is especially true for high-power/current profiles such as HEV-type ones. Aging tests can be designed in such a way that the chamber temperature is controlled, while the battery casing temperature is monitored. Another possibility is to adjust the chamber temperature in real time so that the battery casing temperature is set to a constant. The first method is the most common but makes the aging analysis more complicated since the cell temperature tends to increase as it ages (as the result of a cell impedance increase); the second option is definitely more advanced, but requires more sophisticated equipment.

4.1.2 Nonintrusive Cell Characterization Techniques

The storage and cycling protocols discussed in the above subsection are interrupted at some time interval to characterize the cell, so as to evaluate its SOH. Just like for the aging tests themselves, a variety of different tests can be performed, the selection of which is guided by the type of aging analysis that is sought. Characterization tests should be designed so that they produce little or no cell degradation compared with the aging test itself. There are two ways to address that. First, the cell is usually characterized in temperature conditions that are virtually not damaging for the cell. 25 °C is the most common temperature for characterization and also constitutes a reference temperature for all tests. Second, the characterization test duration is usually taken to a minimum in order to maximize the aging test time.

There are two types of characterization tests: those aimed to assess the remaining energy or capacity of the cell, and the others aimed to determine its remaining power capability. Capacity measurements are usually done by charging/discharging the cell at low rate. In order to save time, a constant current/constant voltage (CCCV) procedure may be used instead of the low-current cycling. In such a measurement, the cell is charged (discharged) galvanostatically up (down) to a cutoff voltage, after which that voltage is held constant until the current decays to a minimum value. The total capacity approaches that measured by charging/discharging the cell at that minimum current value. Another alternative is to use a charge/discharge made up of the succession of decreasing current events [52]. Basically, each time the cutoff potential is reached at a specified current value, a smaller current is used after a quick relaxation period, until the potential reaches the cutoff value again. By cumulating capacities, it is possible to determine the capacity of the cell at each rate, which provides the so-called rate capability of the cell. Charge/discharge curves at different C-rates are much useful to adjust input parameters of physics-based models such as the rate constant k^0 and the solid diffusion coefficient D_s.

The low-rate capacity depends upon the internal balancing of the cell (Fig. 3), and therefore it is impacted by cyclable lithium loss and active material loss(es). An analysis of the differential capacity (dQ/dV vs. V) and differential voltage (dV/dQ vs. Q) curves at low current is helpful to decipher the amount of capacity loss due to cyclable lithium and that due to active material loss [53–57].

Cyclable lithium loss results from side reactions at both electrodes. As it is illustrated in Fig. 7, if the charge consumed in reductive side reactions at anode exactly matches that involved in oxidative side reactions at cathode, no cyclable lithium loss is experienced. Interestingly, this is what happens if a side reaction product at one electrode reacts with another side reaction at the opposite electrode (i.e., shuttle mechanism). However, it is commonplace that the charge consumed in anode side reactions exceeds that consumed at cathode, because of the SEI formation and growth. For modeling battery aging, it is of interest to experimentally determine amounts of charge involved in side reactions at both electrodes. For the case of constant-current aging tests, they can be experimentally accessed from accurate measurement of the end-of-charge (EOC) and end-of-discharge (EOD) capacity slippages [43, 58] (Fig. 13).

In the absence of active material loss, the cumulative EOC slippage is a direct reading of the charge involved in oxidative side reactions at cathode, whereas the cumulative EOD slippage corresponds to the charge involved in reductive side reactions at anode. The limitation of the method is that a precise coulometer is needed to measure these slippages; otherwise, errors cumulate and results are not meaningful [59]. Performing aging tests of long duration with this type of sophisticated equipment has necessarily some cost implications.

In the case of storage experiments under open circuit, it is also possible to measure charges involved in anode and cathode side reactions. As illustrated in Fig. 7, the irreversible, reversible, and total capacity losses at characterization #i (i.e., $\Delta Q_{irr,i}$, $\Delta Q_{rev,i}$, and $\Delta Q_{tot,i}$, respectively) are calculated according to

Fig. 13 Schematics depicting the end-of-discharge (ΔQ_d) and end-of-charge (ΔQ_c) capacity slippages of a Li-ion cell (Adapted from [58])

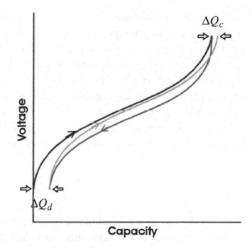

$$\Delta Q_{\text{tot},i} = [Q_{2,i-1} - (1 - SOC)Q_0] - Q_{1,i}$$
$$\Delta Q_{\text{irr},i} = Q_{2,i-1} - Q_{2,i} \qquad (21)$$
$$\Delta Q_{\text{rev},i} = \Delta Q_{\text{tot},i} - \Delta Q_{\text{irr},i}$$

with

$Q_{1,i}$ Residual capacity (First discharge after the storage period) at characterization #i (Ah),

$Q_{2,i}$ Capacity measured at the second discharge at characterization #i (Ah),

Q_0 Reference capacity used to reSOC the cell at the end of characterization #$i - 1$ (Either the initial battery capacity at beginning of life or $Q_{2,i-1}$ are used depending upon the aging test protocols.) (Ah), and

SOC Storage SOC.

Because the charge involved in side reactions at anode generally exceeds that involved at cathode, as mentioned above, the cumulated total capacity loss ($\sum_i \Delta Q_{\text{tot},i}$) is a reading of the charge involved in side reactions at anode, whereas $\sum_i \Delta Q_{\text{rev},i}$ is a reading of the charge involved in side reaction at cathode [56, 57, 60].

Let us now move on to methods for assessing the cell power capability. Electrochemical impedance spectroscopy (EIS) is a technique commonly included in the characterization protocols to analyze the cell dynamics. Because battery operation is transient, EIS is generally performed at open circuit (i.e., there is no direct current component) and at different SOCs of the battery. EIS allows for an in-depth analysis of the various impedance contributions by sorting them out according to their time constants. The evolution of impedance spectra over aging usually provides insight on the origin of the impedance increase in the cell (e.g., interfacial phenomena vs. bulk transport phenomena). Note that although equivalent circuit models are popular to analyze EIS data, it is possible to simulate impedance spectra directly from a physics-based model such as *Dualfoil*, without resorting to an equivalent circuit analog [61]. Data analysis with a physics-based model allows in principle for adjusting input parameters such as solid diffusion coefficient D_s in a more accurate way than with methods like in Ref. [62] that are subject to a number of approximations [61].

Although EIS is a powerful technique, standard pulse tests also prove attractive to evaluate the power capability of the cell. They are indeed popular within the electrical engineering community. The peak power test profile reported in the USABC consortium manual of the US DOE is such a typical test [51]. The resistance at 30 s in Fig. 14 is calculated as

$$R_{30s} = \frac{\Delta V}{\Delta I} = \frac{V_1 - V_2}{I_1 - I_2}$$ (22)

with V_1, V_2, I_1, and I_2 as defined in the figure.

Other techniques are sometime included in nonintrusive characterizations, such as the potentiostatic and galvanostatic intermittent titration techniques (PITT and GITT), which bring complementary information to the other above-mentioned techniques. GITT is selected whenever an accurate measurement of the open-circuit potential of the cell is required. Note that EIS and pulse tests, which are frequently performed at different SOCs, as mentioned above, can easily be combined with the GITT technique. Pulse tests and EIS are then repeated after each rest period of the GITT. The duration of the rest period results from a tradeoff between the time required for the cell to reach equilibrium and the total characterization time that should not be too long.

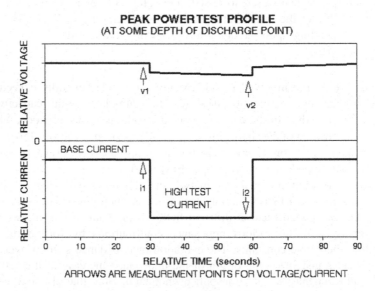

Fig. 14 Example of peak power test profile (at some depth of discharge point) provided in the USABC manual for peak power calculations [51]

4.1.3 Intrusive Analysis

Intrusive cell analysis (also called postmortem for "aged" cells) is generally performed once the aging test is over. A number of characterization techniques are implemented on some cell components with the primary goal of identifying the different aging phenomena. The intrusive analysis is complementary to the non-intrusive characterization tests detailed in the above subsection. In this work, we restrict our description to the most commonly used techniques. The reader is advised to consult Ref. [63] for a dedicated review.

Typically, the Li-ion cell (aged or not) is deeply discharged and taken apart in an Ar-filled glove box. The dismantling procedure depends upon the type of cells (e.g., pouch, aluminum casing, steel casing) and may require custom-designed tools to avoid internal short circuits. The two electrodes and the separator are taken apart and samples of these are rinsed in dimethylcarbonate (DMC) in order to remove the remaining Li salt. Multiple rinsing steps in DMC are generally required, their number and duration being adjusted to remove the majority of the lithium salt while avoiding the partial dissolution of some SEI components. Once vacuum dried, the electrode samples are ready for characterization. The dismantling step, although mainly visual, is crucial since it allows for a qualitative assessment of the general aspect of the electrodes and how uniform it is over their surface. It is common to visually detect a weakening of the mechanical properties of the electrodes as outlined by the presence of cracks in the electrode laminate and/or a delamination of the laminate from the current collector. In other instances, electrodes exhibit whitish stains at some locations that usually arise from accumulated decomposition products in the pores or at the surface. The visual examination commonly helps guiding the type of analysis to be done on the cell components, and to properly select the samples so that they are altogether the representative of the overall electrodes.

To begin with, a number of simple methods turn out to be enlightening. For instance, the difference in weight between an aged and a pristine electrode sample allows determining the mass of solid products (e.g., SEI compounds) that accumulated in the electrode upon aging. The change in electrode thickness is another interesting aging metrics; aged electrodes are generally thicker than their pristine counterparts as the result of the accumulation of solid degradation products. Besides, electrode thickness allows determining the volume fraction of pores in the electrodes, when combined with pycnometric measurements.

Scanning electron microscopy (SEM) is used to study the morphology of the electrodes at a micron scale or less. The microscope is generally coupled with energy-dispersive X-ray analysis (EDX), and hence it is possible to correlate a change in morphology with a change in composition at a local scale. For example, Fig. 15 presents the evolution in morphology of graphite electrodes before (a) and after calendar aging at full charge and at 45 °C (b) and 60 °C (c). Micrographs unambiguously show that graphite particles are covered with a layer which is more pronounced at higher aging temperature. This layer is made up of electrolyte decomposition products forming the SEI. Elemental analysis (averaged out over the electrode surface) reveals a larger content in O and F for aged electrodes.

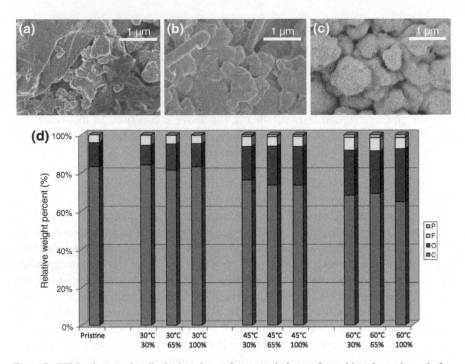

Fig. 15 SEM micrographs displaying the surface morphology of graphite electrodes **a** before aging, **b** after calendar aging at 100 % SOC and 45 °C, **c** after calendar aging at 100 % SOC and 60 °C. **d** Corresponding EDX analyses at three different temperatures and three different SOCs. Data from the Simcal project [64]

Furthermore, the amount of O and F is larger for electrodes aged in severe storage conditions (i.e., high SOC and T).

Elemental analyses with regular techniques, such as the inductive coupled plasma-optical emission spectroscopy (ICPOES), allow for the quantitative analysis of many elements. To this end, the sample to be analyzed is dissolved in an aqueous solution of nitric acid. For instance, this method allows the titration of the total lithium content in a graphite electrode sample, from which the amount of inactive lithium is derived by subtracting the amount of active lithium. The amount of inactive lithium, once scaled up to the overall electrode, should compare well with the total charge consumed in side reactions at that same electrode (i.e., $\Delta Q_{irr} + \Delta Q_{rev}$ at Fig. 7). Such a cross check between techniques is a good way to assess whether some electrolyte degradation products are soluble in the electrolyte or if all reduction products end up as solid SEI products. The analysis of contaminants is of interest as well; for instance, transition metal elements resulting from cathode material dissolution and ending up at the anode side can be quantified.

The porous volume of the electrodes and of the separator is generally determined using He pycnometry. A sample of known surface area and thickness is introduced into the pycnometer. The volume fraction of pores in the electrode/separator can then be derived. The accumulation of solid degradation products in the electrode usually leads to a decrease of the porosity, and consequently of the cell performance [50].

The analysis of the electrodes by X-ray diffraction (XRD) allows for detecting crystalline impurity phases, in addition to an assessment of the active material itself (e.g., Li content, degree of crystallinity). When the active material inserts/deinserts Li according to a two-phase process, a Rietveld analysis of the XRD diagrams allows for phase quantification and Li content is then derived. For single-phase materials, lattice parameters change with the Li content, and therefore it can be evaluated as well. Figure 16 presents an example of XRD analysis for LFP electrodes (two-phase material) recovered from calendar-aged cells. As the aging temperature increases, the Li content in the discharged cathode material recovered from the cell decreases. This arises from a change in the internal balancing of the graphite/LFP cells during aging. Besides assessing composition, XRD is also a good tool to examine the loss in crystallinity of an electrode material after aging. Methods such as that by Williamson and Hall [65] allow deciphering the loss of crystallinity due to a change in crystallite size and that due to microstrains.

Characterization methods directly applicable to solid and soluble degradation products can help unravel some aging mechanisms. Solid products are typically analyzed by methods such as infrared spectroscopy, X-ray photoelectron spectroscopy, etc. Soluble products are analyzed by methods such as liquid or gas chromatography coupled with mass spectrometry.

The individual electrode characterization by electrochemical methods is probably the most useful intrusive characterization tool to analyze aging. Samples of harvested electrodes from Li-ion cells are double-side coated. Hence, the coating at one side must be removed from the current collector before the electrode sample can be used in a lab cell. This is usually done by rubbing the coating at one side with a solvent that dissolves the electrode binder (e.g., N-methyl pyrrolidone for

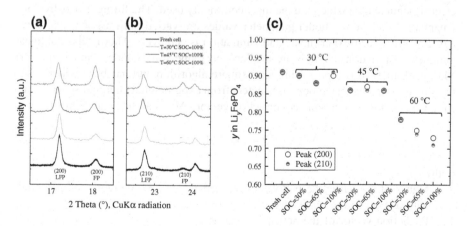

Fig. 16 a XRD patterns of calendar-aged cathodes at different temperatures and **b** Li content derived from a quantitative analysis of the XRD patterns [66]

PVdF binder). Once done, electrode samples are punched and are assembled vs. a Li metal foil (Li is in large excess), in two- or three-electrode cells (i.e., without or with a reference electrode, respectively). The residual capacity measurement done by charging or discharging the electrode under low rate is an electrochemical titration method of the initial active Li content in the electrode (just before cell opening) and should compare with the Li content derived from XRD. The difference between the overall Li content measured with methods such as elemental analysis and the active Li content determined from XRD or electrochemical titration gives access to the inactive Li content. The capacity of the electrode, measured by charge/discharge at low rate, is a direct reading of the amount of active material in the electrode, and thus allows for quantifying active material losses. In addition, it is of interest to run electrochemical tests to evaluate how the electrode reacts under high current draws. Methods like EIS, pulse tests, and rate capability tests, which are described in Sect. 4.1.2 for the Li-ion cells, are of interest as well on the half cells. In particular, it is frequent that EIS is carried out on individual electrodes to evaluate which is the one that experiences the higher impedance rise over aging. Furthermore, impedance analysis with a physics-based model gives access to model parameters that would be difficult to determine accurately from time-domain measurements [61].

4.2   "Snapshot" Analysis with the Aging-Free Model

Analysis of data from the nonintrusive characterization tests with a performance model such as those introduced at Sect. 3.1 allows in principle for getting insight into aging phenomena. The analysis usually consists of fitting some model parameters so that simulations match the experimental electrochemical data of the characterization test. Charge/discharge curves at different C-rates are the type of electrochemical data which are the most commonly used. The fitting is usually done "manually," i.e., some model parameter values are varied until a good qualitative agreement between experiments and simulation is achieved. Alternately, nonlinear parameter estimation methods are available to refine some model parameters. The Marquardt method is a promising least-square algorithm and has been widely used to estimate some battery model parameters from experimental charge/discharge data [67]. Calculation of a parameter correction vector ($\Delta\theta$) is at the heart of this method

$$\Delta\theta = \left(J^T J + \lambda I\right)^{-1} J^T (Y^* - Y), \tag{23}$$

with

J Matrix of the partial derivatives of the dependent variable (e.g., cell potential) with respect to the parameters to be estimated,

Y Prediction vector of the dependent variable,

Y^* Experimental vector of the dependent variable,
λ Step size of the correction factor, and
I Identity matrix.

In Eq. 23, T and -1 are used to represent the transpose and inverse of a matrix. In this algorithm, the convergence is highly sensitive to the accurate calculation of J. The sensitivity approach is frequently used in the place of finite-difference method to obtain accurate J. In this approach, for every fitting parameter, a set of sensitivity equations is generated by taking the partial derivative with respect to the fitting parameter on both sides of the model governing equations. J is then directly obtained upon solution of the governing and sensitivity equations [68].

The evolution of model parameters during aging is the output of this type of analysis, and because the model is based on physics, it enables in principle to identify the underlying aging degradation modes, and at least their effects (e.g., cyclable lithium loss, active material loss).

Figure 17a provides an example of such an analysis. Discharge curves of a carbon/Li(Ni,Co)O$_2$ cell at different cycle numbers are presented. Markers and solid lines stand for experimental data and simulations done with the single-particle model, respectively. Only three-model parameters were modified to fit the overall set of discharge curves: $x_{0,neg}$ and $x_{0,pos}$, which represent the lithium contents in the anode and cathode at the end of charge, respectively, and ε_{pos}, which denotes the volume fraction of active material in the cathode. The decrease in ε_{pos} over cycle number indicates that the cathode experiences active material loss during repeated cycling. $x_{0,neg}$ and $x_{0,pos}$ directly relate to the internal balancing of the cell, which changes because of both cyclable lithium loss and active material loss (Fig. 4).

In Fig. 17, cycling data at constant current are directly used as input data for fitting with the single-particle model. Note, however, that this type of analysis is not restricted to cycling tests as aging protocol. Nonintrusive characterization tests

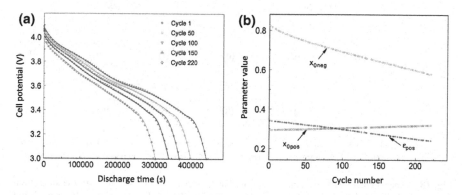

Fig. 17 **a** Discharge curves at different cycle numbers (*Markers* Experiments, *Solid lines* Simulations) for a carbon/Li(Co,Ni)O$_2$ cell and **b** evolution of some model parameters resulting from data fitting with the single-particle model [69]

carried out periodically over the course of the aging test (as described in Sect. 4.1.2) provide the input data to perform this type of analysis. Examples can be found in Refs. [40, 70] where this type of analysis is done on cells aged under storage. In general, the type of analysis described above is coupled with intrusive cell characterizations that are performed once the aging test is completed (Sect. 4.1.3) [40].

Differential capacity (dQ/dV vs. V) and differential voltage (dV/dQ vs. Q) analyses are very helpful not only to elucidate the aging modes but also are precious data to be used for the refinement of some model parameters. Differential plots can be easily created by numerical differentiation of the constant-current charge/discharge data. In differential capacity plots, the phase transitions in active materials show up as sharp peaks, and any change in their position or intensity is a sign of aging. Differential voltage analysis is equally insightful. In Fig. 18a, b, the constant-current discharge profiles at C/10 recorded during the characterization periods are presented for the cycle aging of a 26650 LFP/graphite cells at 25 and 45 °C, respectively [56]. Parameters of an aging-free single-particle model of the cell were manually refined to fit the experimental data (markers) in Fig. 18 [40]. The differential voltage plots have been also simulated and compared to the experiments (Fig. 18c–n). Here, the position and intensity of the peaks are very sensitive to the stoichiometry (x_{max}) and volume fraction (ε_a) of the graphite electrode at the beginning of discharge. Conversely, in the V–Q plots, visual

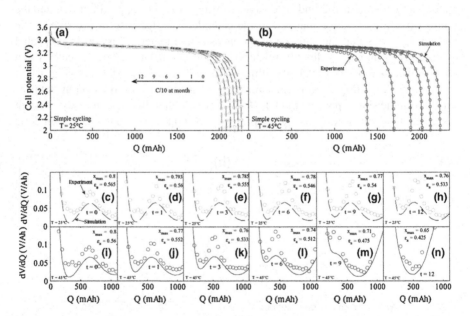

Fig. 18 Experimental C/10 constant-current discharge (*circle markers*) along with the simulation results (*dashed* and *dotted lines*) for the cycle aging of the 26650 LFP/graphite cells at **a** 25 and **b** 45 °C. The corresponding differential voltage curves of the experimental and simulated C/10 discharge profiles at **c–h** 25 °C and **i–n** 45 °C [40]

detection of the simulation/experiment sensitivities to x_{max} and ε_a is not trivial and lends ambiguity to the adjusted values of these model parameters. In Fig. 18, the analysis unequivocally reveals that cycle aging induces both loss of cyclable lithium and graphite active material at both 25 and 45 °C.

At this stage, we have presented examples of "snapshot" analyses that have focused on the time frame of the aging tests only. What is going on beyond aging experiments has not been discussed. Based on simple model parameter extrapolation using empirical functions, it is in principle possible to determine what the cell performance would be at times extending beyond the experiments. The approach is essentially similar to what is frequently done in the literature for capacity and internal resistance projection. This is also similar to the procedure used for equivalent circuits and other empirical models. The advantage offered by carrying such an analysis with a physics-based model is that it is possible to predict the occurrence of new degradation phenomena. For instance, a high extrapolated value of the Li content in the anode at full charge would indicate possible Li plating. Conversely, a lower-than-expected value of the extrapolated Li content in a LCO cathode at full charge would suggest active material degradation because of the irreversible phase transition to the monoclinic form (Sect. 2.2). Empirical models, on the other hand, do not provide this type of information.

An empirical extrapolation of physics-based parameters is presented in Fig. 19 for aging data on a graphite/LCO cell [71]. Only two parameters of a sandwich-type model similar to that introduced in Sect. 3.1 were found to vary during aging, namely the SOC of the anode at full charge θ_n and the anode SEI resistance R_f. The empirical function which was selected for both parameters is a linear variation with the square root of the cycle number. Here, a function of the cycle number was selected, but functions of other variables such as charge throughput or time are sometime employed. For storage experiments, time is the only variable of choice.

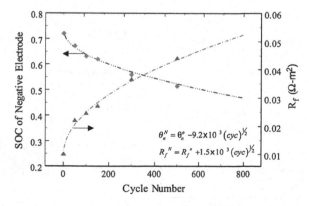

Fig. 19 Evolution of input parameters of sandwich-type model during cycling of 18650 graphite/LCO Sony cells, and empirical fits used to extrapolate the parameter values at larger cycle number [71]

4.3 Analysis with the Aging Model

The most sophisticated approach, which is extrapolation-free in theory, involves including the degradation phenomena directly into the physics-based model, in order to construct a physics-based *aging* model (Sect. 3.2). This type of model can be thought as a numerical cell that can be "aged" by running a numerical aging test to it. Any aging test can in principle be applied to the numerical cell, and importantly, the test can differ from those studied experimentally to calibrate the model. Still, it must lie in conditions that are similar to those investigated experimentally. For instance, if the 20–50 °C temperature range was studied experimentally, simulating an aging test where the temperature exceeds 50 °C might not be meaningful. The same is true, for example, for current and SOC windows. However, it is doable to mix different conditions that were explored in independent aging tests. For instance, an EV profile can be constructed by combining profiles featuring an EV either under operation, under parking, or being recharged. It is still meaningful to build up such a profile, although experiments which served for model calibration were either full-time storage or full-time cycling. The made-up profile is simply run to the aging model without the need for a "reconstruction" that is required in empirical approaches like those derived from mechanical fatigue [72]. In order to assess the SOH of the battery over the simulated aging test, characterization tests are included in the simulated profile, just like it is done in actual experiments.

The main limitation of the approach is to build an aging model which is comprehensive enough so that simulation results beyond the time frame of the experiments are meaningful. As touched upon in the introduction, this is still challenging because not only is a comprehensive identification of all aging phenomena in the cell required, but also these phenomena need to be cast into a meaningful mathematical form for being included in the aging model. In some cases, the full validation of the model within the time frame of the experiments itself proves tricky. This is outlined in Fig. 20a where experimental and simulated discharge capacities

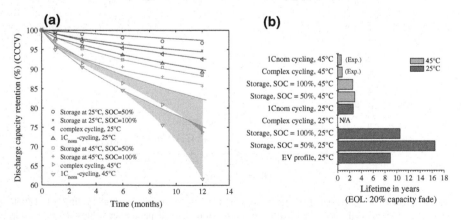

Fig. 20 **a** Aging model validation for a 26650 graphite/LiFePO$_4$ and **b** prediction of the cell lifetime for different aging conditions with a physics-based aging model [43]

of 26650 graphite/LFP cells for eight different aging test conditions are represented with markers and solid lines, respectively [43]. The aging model captures the experimental variations pretty well, except for the two conditions which experienced the largest capacity losses. This discrepancy means that model validation is not complete, at least for large capacity fade. There is probably an additional aging phenomenon that is not included in the aging model and that kicks in whenever capacity fade becomes substantial. Nevertheless, if we assume the model to be valid at low capacity fade (typically <20 %), it can be used for life prediction. A simplified EV profile is run to the aging model up to a decrease of 20 % of the initial capacity which is selected as the end-of-life (EOL) condition. Results of the simulations are provided in Fig. 20b for all eight conditions studied experimentally as well as for the simplified EV profile. A life time of ~9 years is forecast for this type of cells, whereas experiments to calibrate the model did not go over a year.

The aging-model parameters refined experimentally for a given SOC window might be scaled and used for prediction of aging across other ranges of SOC. In Ref. [47], successful derivation and application of such a scaling factor was demonstrated for mechanical fatigue (i.e., fracture and isolation) of electrodes under cycle aging. In this work, $\Delta SOC \times SOC_{mean}$ was theoretically identified as the key

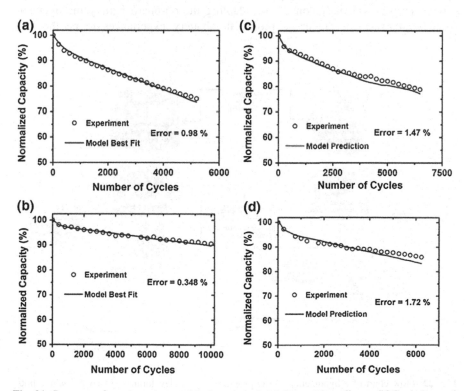

Fig. 21 Parameter fit to the aging model of a prismatic LMO/graphite cell over SOC windows of **a** 100–25 % and **b** 75–65 %, together with the model predictions for SOC ranges of **c** 100–50 % and **d** 75–25 % [47]

measure of charge–discharge cycle fatigue. Figure 21a, b shows the fits of model to the aging experimental data for the prismatic graphite/LMO cell cycled at SOC window 100–25 and 75–65 %, respectively. The aging parameters are scaled for the 100–50 and 75–25 % and the model predictions show promising matches with experiments as presented in Fig. 21 c, d, respectively.

The examples mentioned above clearly demonstrate the advantage of physics-based approach in terms of time/cost savings.

5 Outlook of Physics-Based Aging Modeling

So far, the methodology introduced in Fig. 2 and detailed in Sects. 4.2 and 4.3 consisted in life prediction of selected cells under specific usage conditions of an application, using physics-based (aging) models. To go further, we present in Fig. 22 how this methodology could be implemented in a hypothetical R&D program aiming to develop batteries for an EV. In this research scheme, the physics-based model is used in order to maximize the battery lifetime through an optimization of both the driving profile and of the cell design. Optimization of the driving profile essentially consists of adapting the on-board battery management strategy in order for the usage profile the battery experiences to be the less

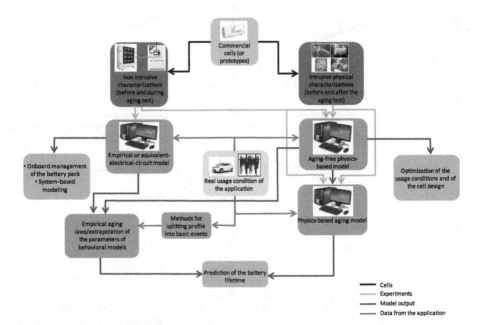

Fig. 22 Flow chart of a hypothetical R&D program for the development of an EV where both empirical and physics-based models are used for performance and aging assessment and improvement

damaging to its performance over time. Toward this goal, the physics-based aging model allows for quick simulations of the battery life under different input usage profiles, as detailed in Sect. 4.3. Optimization of the cell design is more speculative and would consist in providing feedback to the cell manufacturer based on the results of the aging analysis with the physics-based model. Let us provide an example here. Assuming the primary reason for capacity fade is the loss of cyclable lithium because of side reactions leading to SEI growth at the anode, one could provide directions to the cell manufacturer to increase the charge involved in side reactions at cathode in order to counterbalance the charge consumed at anode. Because the extent of side reactions depends upon operating conditions such as current and temperature, guidance on cell design provided to the cell maker should account for the typical operating conditions experienced by the cell, and hence the tuning of side reactions at cathode should result from an optimization with the aging model in realistic usage conditions.

Another application of physics-based aging models is to guide the development of empirical methodologies for battery aging studies. Indeed, large aging test programs require lots of resource (e.g., channels on large-current cyclers, temperature chambers) and for long amounts of time, and there is also human resource needed to supervise and monitor the tests. It has direct implications in terms of cost. Physics-based aging models can be used as numerical batteries and quick numerical experimentation can be carried out in order to help design the aging test program and make it more robust. Such a methodology had been applied successfully to develop a test program based on concepts of mechanical fatigue; a physics-based aging model of a graphite/LCO 18650 cell with a growing SEI at anode as a single aging source was used for numerical experimentation [73, 74].

To conclude, although physics-based modeling of battery aging is still at a somewhat early stage of development as of today, it is expected to rise with a similar pace as that of Li-ion batteries for long-lasting applications, such as electric transportation and grid storage. In this journey, commercial physics-based models of Li-ion batteries [75, 76] have started R&D developments to include aging phenomena in their codes. The success of the approach will greatly depend on a joint effort between experimentalists advancing the understanding of aging phenomena and engineers developing the modeling codes.

References

1. Nagaura T, Tazawa K (1990) Lithium-ion rechargeable battery. Prog Batteries Solar Cells 9:209
2. Amatucci G, Tarascon J-M (2002) Optimization of insertion compounds such as $LiMn_2O_4$ for Li-ion batteries. J Electrochem Soc 149(12):K31
3. Numata T, Amemiya C, Kumeuchi T, Shirakata M, Yonezawa M (2001) Advantages of blending $LiNi_{0.8}Co_{0.2}O2$ into $Li_{1+x}Mn_{2-x}O4$ cathodes. J Power Sources 97–98:358
4. Chikkannanavar SB, Bernardi DM, Liu L (2014) A review of blended cathode materials for use in Li-ion batteries. J Power Sources 248:91–100

5. Tran HY, Taubert C, Fleischhammer M, Axmann P, Kuppers I, Wohlfart-Mehrens M (2011) $LiMn_2O_4$ spinel/$LiNi_{0.8}Co_{0.15}Al_{0.05}O_2$ blends as cathode materials for lithium-ion batteries. J Electrochem Soc 158:A556–A561

6. Srinivasan V (2008) Batteries for vehicular applications. In: AIP conference proceedings, vol 1044. AIP Publishing, pp 283–296

7. http://www.uscar.org/ (1996)

8. Newman J, Hoertz PG, Bonino CA, Trainham JA (2012) Review: an economic perspective on liquid solar fuels. J Electrochem Soc 159(10):A1722–A1729

9. Darling R, Newman J (1998) Modeling side reactions in composite $Li_yMn_2O_4$ electrodes. J Electrochem Soc 145(3):990–998

10. Arora P, Doyle M, White RE (1999) Mathematical modeling of the lithium deposition overcharge reaction in lithium-ion batteries using carbon-based negative electrodes. J Electrochem Soc 146(10):3543–3553

11. Christensen J, Newman J (2004) A mathematical model for the lithium-ion negative electrode solid electrolyte interphase. J Electrochem Soc 151(11):A1977–A1988

12. Ramadass P, Haran B, Gomadam PM, White R, Popov BN (2004) Development of first principles capacity fade model for Li-ion cells. J Electrochem Soc 151(2):A196–A203

13. Safari M, Morcrette M, Teyssot A, Delacourt C (2009) Multimodal physics-based aging model for life prediction of Li-ion batteries. J Electrochem Soc 156(3):A145–A153

14. Arora P, White RE, Doyle M (1998) Capacity fade mechanisms and side reactions in lithium-ion batteries. J Electrochem Soc 145(10):3647–3667

15. Vetter J, Novák P, Wagner MR, Veit C, Möller K-C, Besenhard JO, Winter M, Wohlfahrt-Mehrens M, Vogler C, Hammouche A (2005) Aging mechanisms in lithium-ion batteries. J Power Sources 147(1–2):269–281

16. Christensen J, Newman J (2005) Cyclable lithium and capacity loss in Li-ion ells. J Electrochem Soc 152(4):A818–A829

17. Peled E (1979) Electrochemical behavior of alkali and alkaline earth metals in nonaqueous battery systems – the solid electrolyte interphase model. J Electrochem Soc 126(12):2047–2051

18. Inaba M, Tomiyasu H, Tasaka A, Jeong SK, Ogumi Z (2004) Atomic force microscopy study on the stability of a surface film formed on a graphite negative electrode at elevated temperatures. Langmuir 20:1348–1355

19. Delacourt C, Kwong A, Liu X, Qiao R, Yang WL, Lu P, Harris SJ, Srinivasan V (2013) Effect of manganese contamination on the solid-electrolyte-interphase properties in Li-ion batteries. J Electrochem Soc 160(8):A1099–A1107

20. Xiao X, Liu Z, Baggetto L, Veith GM, More KL, Unocic RR (2014) Unraveling manganese dissolution/deposition mechanisms on the negative electrode in lithium ion batteries. Phys Chem Chem Phys 16:10398–10402

21. Waldmann T, Wilka M, Kasper M, Fleischhammer M, Wohlfahrt-Mehrens M (2014) Temperature dependent ageing mechanisms in lithium-ion batteries – a post-mortem study. J Power Sources 262:129–135

22. Tang M, Albertus P, Newman J (2009) Two-dimensional modeling of lithium deposition during cell charging. J Electrochem Soc 156(5):A390–A399

23. Deshpande R, Verbrugge M, Cheng Y-T, Wang J, Liu P (2012) Battery cycle life prediction with coupled chemical degradation and fatigue mechanics. J Electrochem Soc 159(10): A1730–A1738

24. Andersson AM, Edstrom K, Rao N, Wendsjo A (1999) Temperature dependence of the passivation layer on graphite. J Power Sources 81–82:286–290

25. Aurbach D, Koltypin M, Teller H (2002) In situ AFM imaging of surface phenomena on composite graphite electrodes during lithium insertion. Langmuir 18:9000–9009

26. Svens P, Eriksson R, Hansson J, Behm M, Gustafsson T, Lindbergh G (2014) Analysis of aging of commercial composite metal oxide – $Li_4Ti_5O_{12}$ battery cells. J Power Sources 270:131–141

27. Castaing R, Reynier Y, Dupre N, Schleich D, Jouanneau Si Larbi S, Guyomard, D, Moreau P (2014) Degradation diagnosis of aged $Li_4Ti_5O_{12}$/$LiFePO_4$ batteries. J Power Sources 267:744–752

28. Wohlfahrt-Mehrens M, Vogler C, Garche J (2004) Aging mechanisms of lithium cathode materials. J Power Sources 127:58–64
29. Edstrom K, Gustafsson T, Thomas JO (2004) The cathode-electrolyte interface in the li-ion battery. Electrochim Acta 50:397–403
30. Aurbach D, Markovsky B, Salitra G, Markevich E, Talyossef Y, Koltypin M, Nazar L, Ellis B, Kovacheva D (2007) Review on electrode-electrolyte solution interfaces related to cathode materials for Li-ion batteries. J Power Sources 165:491–499
31. Reimers JN, Dahn JR (1992) Electrochemical and in situ x-ray diffraction studies of lithium intercalation in Li_xCoO_2. J Electrochem Soc 139(8):2091–2096
32. Blyr A, Sigala C, Amatucci G, Guyomard D, Chabre Y, Tarascon J-M (1998) Self-discharge of $LiMn_2O_4$/C Li-ion cells in their discharged state: understanding by means of three-electrode measurements. J Electrochem Soc 145(1):194–209
33. Kim D, Park S, Chae OB, Ryu JH, Yin R, Kim Y-U, Oh S (2012) Re-deposition of manganese species on spinel LiMn2O4 electrode after Mn dissolution. J Electrochem Soc 159(3):A193–A197
34. Koltypin M, Aurbach D, Nazar L, Ellis B (2007) On the stability of LiFePO4 olivine cathodes under various conditions (electrolyte solutions, temperatures). Electrochem Solid State Lett 10 (2):A40–A44
35. http://www.cchem.berkeley.edu/jsngrp/
36. Thomas-Alyea KE, Darling RM, Newman J (2002) In: Schalkwijk WV, Scrosati B (eds) Mathematical modeling of Lithium batteries. Advances in lithium-ion batteries, chapter 12. KluwerAcademic/Plenum Publishers
37. Newman JS, Thomas-Alyea KE (2004) Electrochemical systems. Wiley Interscience
38. Atlung S, West K, Jacobsen T (1979) Dynamic aspects of solid solution cathodes for electrochemical power sources. J Electrochem Soc 129:1311
39. Santhanagopalan S, Guo Q, Ramadass P, White RE (2006) Review of models for predicting the cycling performance of lithium ion batteries. J Power Sources 156:620–628
40. Safari M, Delacourt C (2011) Simulation-based analysis of aging phenomena in a commercial graphite/LiFePO4 cell. J Electrochem Soc 158(12):A1436–A1447
41. Ploehn HJ, Ramadass P, White RE (2004) Solvent diffusion model for aging of lithium-ion battery cells. J Electrochem Soc 151(3):A456–A462
42. Broussely M, Herreyre S, Biensan P, Kasztejna P, Nechev K, Staniewicz RJ (2001) Aging mechanisms inn Li-ion cells and calendar life predictions. J Power Sources 97–98:13–21
43. Delacourt C, Safari M (2012) Life simulation of a graphite/LiFePO4 cell under cycling and storage. J Electrochem Soc 159(8):A1283
44. Perkins RD, Randall AV, Zhang X, Plett GL (2012) Controls oriented reduced order modeling of lithium deposition on overcharge. J Power Sources 209:318–325
45. Christensen J, Newman J (2006) Stress generation and fracture in lithium insertion materials. J Solid State Electrochem 10:293–319
46. Verbrugge MW, Cheng Y-T (2009) Stress and strain-energy distributions within diffusion-controlled insertion-electrode particles subjected to periodic potential excitations. J Electrochem Soc 156(11):A927–A937
47. Narayanrao R, Joglekar MM, Inguva S (2013) A phenomenological degradation model for cycling aging of lithium ion cell materials. J Electrochem Soc 160(1):A125–A137
48. Dai Y, Cai L, White RE (2013) Capacity fade model for spinel $LiMn_2O_4$ electrode. J Electrochem Soc 160(1):A182–A190
49. Delacourt C (2013) Modeling Li-ion batteries with electrolyte additives or contaminants. J Electrochem Soc 160(11):A1997–A2004
50. Sikha G, Popov BN, White RE (2004) Effect of porosity on the capacity fade of a lithium-ion battery theory. J Electrochem Soc 151(7):A1104–A1114
51. http://www.uscar.org/guest/publications.php (1996)
52. Doyle M, Newman J, Reimers J (1994) A quick method of measuring the capacity versus discharge rate for a dual lithium-ion insertion cell undergoing cycling. J Power Sources 52:211–216

53. Bloom I, Jansen AN, Abraham DP, Knuth J, Jones SA, Battaglia VS, Henriksen GL (2005) Differential voltage analyses of high-power lithium-ion cells, 1. Technique and application. J Power Sources 139(1–2):295–303

54. Bloom I, Christophersen J, Gering K (2005) Differential voltage analyses of high-power lithium-ion cells, 2.applications. J Power Sources 139(1–2):304–313

55. Dubarry M, Svoboda V, Hwu R, Liaw BY (2006) Incremental capacity analysis and close-to-equilibrium OCV measurements to quantify capacity fade in commercial rechargeable lithium batteries. Electrochem. Solid State Lett. 9(10):A454–A457

56. Safari M, Delacourt C (2011) Aging of a commercial graphite/LiFePO$_4$ cell. J Electrochem Soc 158(10):A1123–A1135

57. Kassem M, Bernard J, Revel R, Pelissier S, Duclaud F, Delacourt C (2012) Calendar aging of a graphite/LiFePO$_4$ cell. J Power Sources 208:296–305

58. Smith AJ, Burns JC, Xiong D, Dahn JR (2011) Interpreting high precision coulometry results on Li-ion cells. J Electrochem Soc 158(10):A1136–A1142

59. Smith AJ, Burns JC, Trussler S, Dahn JR (2010) Precision measurements of the coulombic efficiency of lithium-ion batteries and of electrode materials for lithium-ion batteries. J Electrochem Soc 157(2):A196–A202

60. Sinha NN, Smith AJ, Burns JC, Jain G, Eberman KW, Scott E, Gardner JP, Dahn JR (2011) The use of elevated temperature storage experiments to learn about parasitic reactions in wound LiCoO$_2$/graphite cells. J Electrochem Soc 158(11):A1194–A1201

61. Doyle M, Meyers JP, Newman J (2000) Computer simulations of the impedance response of lithium rechargeable batteries. J Electrochem Soc 147(1):99–110

62. Ho C, Raistrick ID, Huggins RA (1980) Application of A-C techniques to the study of lithium diffusion in tungsten trioxide thin films. J Electrochem Soc 127(2):343–349

63. Nagpure SC, Bhushan B, Babu SS (2013) Multi-scale characterization studies of aged Li-ion large format cells for improved performance: an overview. J Electrochem Soc 160(11): A2111–A2154

64. Moreau F, Bernard J, Revel R (2012) Unpublished results, IFP Energies Nouvelles

65. Williamson GK, Hall WH (1953) X-ray line broadening from field aluminium and wolfram. Acta Metall 1:22

66. Kassem M, Delacourt C (2013) Postmortem analysis of calendar-aged graphite/LiFePO$_4$ cells. J Power Sources 235:159–171

67. Santhanagopalan S, Guo Q, White RE (2007) Parameter estimation and model discrimination for a lithium-ion cell. J Electrochem Soc 154(3):A198–A206

68. Guo Q, Sethuraman VA, White RE (2004) Parameter estimates for a PEMFC cathode. J Electrochem Soc 151(7):A983–A993

69. Zhang Q, White RE (2008) Calendar life study of Li-ion pouch cells. part 2: simulation. J Power Sources 179(2):785–792

70. Zhang Q, White RE (2008) Capacity fade analysis of a lithium-ion cell. J Power Sources 179 (2):793–798

71. Ramadass P, Haran B, White R, Popov BN (2003) Mathematical modeling of the capacity fade of Li-ion cells. J. Power Sources 123:230–240

72. Picciano N (2007) Battery aging and characterization of nickel metal hydride and lead acid batteries. Ph.D. dissertation, The Ohio State University, OH, USA

73. Safari M, Morcrette M, Teyssot A, Delacourt C (2010) Life-prediction methods for lithium-ion batteries derived from a fatigue approach: I. introduction: capacity-loss prediction based on damage accumulation. J Electrochem Soc 157(6):A713–A720

74. Safari M, Morcrette M, Teyssot A, Delacourt C (2010) Life-prediction methods for lithium-ion batteries derived from a fatigue approach: II. capacity-loss prediction of batteries subjected to complex current profiles. J Electrochem Soc 157(7):A892–A898

75. http://www.comsol.com/batteries-and-fuel-cells-module

76. http://www.cd-adapco.com/products/battery-design-studio

Fuel Cells and Batteries In Silico Experimentation Through Integrative Multiscale Modeling

Alejandro A. Franco

Abstract Devices for electrochemical energy conversion and storage exist at different levels of development, from the early stages of R&D to mature and deployed technologies. Thanks to the very significant progresses achieved in the field of computational science over the past few decades, multiscale modeling and numerical simulation are emerging as powerful tools for in silico studies of mechanisms and processes in these devices. These innovative approaches allow linking the chemical/microstructural properties of materials and components with their macroscopic efficiency. In combination with dedicated experiments, they can potentially provide tremendous progress in designing and optimizing the next-generation electrochemical cells. This chapter provides a comprehensive overview of the theory and practical aspects of integrative multiscale modeling tools within the context of fuel cells and rechargeable batteries. Additionally, the chapter discusses technical dreams and methodological challenges that computational science is facing today in order to help developing efficient, durable, and low-cost electrochemical energy devices but also to trigger major technological breakthroughs.

A.A. Franco (✉)
Laboratoire de Réactivité et Chimie des Solides, Université de Picardie Jules Verne
and CNRS, 33 Rue Saint Leu, 80039 Amiens Cedex, France
e-mail: alejandro.franco@u-picardie.fr

A.A. Franco
Réseau sur le Stockage Electrochimique de l'Energie (RS2E), FR CNRS 3459,
Amiens Cedex, France

A.A. Franco
ALISTORE European Research Institute, FR CNRS 3104, 80039 Amiens Cedex, France

© Springer-Verlag London 2016   191
A.A. Franco et al. (eds.), *Physical Multiscale Modeling and Numerical
Simulation of Electrochemical Devices for Energy Conversion and Storage,*
Green Energy and Technology, DOI 10.1007/978-1-4471-5677-2_6

1 Introduction

Current challenges facing humanity nowadays include climate change, depletion of fossil resources, and fast increase in energy demand. In order to ensure prosperity of humanity, a worldwide extensive deployment of renewable energies is urgently needed.

Unfortunately, renewable energies are intermittent and often unpredictable. Energy storage technologies are therefore needed to provide flexibility, balance power grids, and improve management efficiency. Rechargeable batteries, such as lithium ion ones, offer a simple and efficient way to store electricity, and the R&D efforts have to date largely focused on automotive and small-scale portable systems [1]. Large-scale power grid storage requires batteries with durability to large numbers of charge/discharge cycles as well as calendar life, high efficiency, an ability to respond rapidly to changes in load or input, and reasonable costs (Fig. 1).

Another approach to store energy consists of using the hydrogen vector. This requires the implementation of appropriate technologies for hydrogen generation, its storage, and its conversion back to electrical energy. Fuel cells, and more particularly hydrogen-feed Polymer Electrolyte Membrane Fuel Cells (PEMFCs), appear nowadays as the most promising of these electrochemical energy conversion technologies [2].

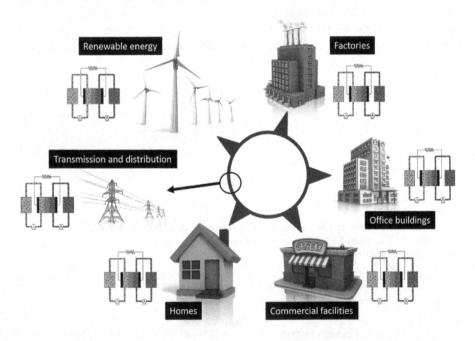

Fig. 1 Possible application domains for stationary batteries in a sustainable energy network

In spite of the excellent prospects in these technologies, massive commercialization of advanced electrochemical devices for power generation in transportation and stationary applications is still not completely guaranteed. For the commercialization of such technologies, the reduction in cost of the materials, the increase of their efficiency and their stability would be the decisive breakthrough. This makes necessary the development of new concepts in the design of advanced materials, operation conditions as well as the fundamental understanding of basic electrochemical processes.

1.1 The Role of Computational Electrochemistry

Electrochemical devices for energy conversion and storage are frequently multiphase systems involving liquids (electrolytes) and solids (electrochemically active surfaces), and sometimes gas (Fig. 2). These devices are also multiphysics systems where a large diversity competing mechanisms occur during their operation, such as electrochemistry, ionic and liquid transport, mechanical stresses, and heat transfer. All these mechanisms are strongly and nonlinearly coupled over various spatial scales, and thus processes at the nano- and microscale can therefore dominantly influence the macroscopic behavior. For example, the materials' spatiotemporal microstructural changes lead into irreversible long-term cell power degradation, and the ways of how aging mechanisms occur are expected to be strongly sensitive to the cell operation mode.

Computational electrochemistry usually refers to the discipline consisting of the use of computational science to simulate specific mechanisms and mechanisms in interaction (processes) in electrochemical systems. The main outcomes of this

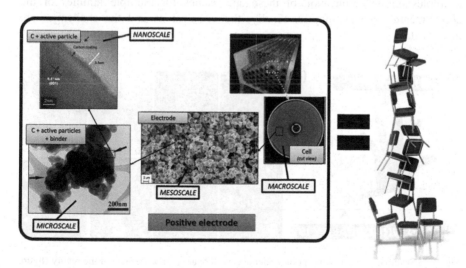

Fig. 2 Schematics of the complexity in a typical Lithium Ion Battery (LIB) electrode (adapted from Ref. [3])

discipline are actually "theories" or "models" which aim to represent the mechanisms and systems under investigation (Fig. 3).

Computational electrochemistry plays multiple roles, for instance:

- "custody" of the knowledge, by supporting the experimentalists in the analysis of their results and by checking the consistency and relevance of the physical knowledge used within this objective. Hence, deep understanding of several individual mechanisms in the cell components, interplays between them and their relative contribution into the global cell response can be developed;
- inspiration source for experimentalists and engineers through the proposal of new "ideas" or "concepts", consisting of innovative materials, designs, or operation conditions;
- predictive tools devoted to answer "what if" questions and consequently providing guidelines for a more efficient experimental work.

Because of the structural complexity and multiphysics character of modern electrochemical devices for energy conversion and storage, interpretation of experimental observations and ultimate cell optimization remain a challenge. Computational electrochemistry provides tools that help in elucidating the efficiency limitations and their location, the degradation and failure mechanisms.

This chapter aims to bring some discussions on the state of the art on multiscale modeling in the field of electrochemical devices for energy storage and conversion, more particularly PEMFCs and rechargeable lithium batteries. Instead of being exhaustive, the chapter focuses on discussing the latest progresses in this discipline and the remaining challenges for it to significantly help on solving many of the technical issues that the energy R&D is facing nowadays.

First, a general introduction is provided where multiscale modeling-related concepts and methods are presented. Then, the chapter provides some illustrative examples on the application of these approaches for multiple families of the above-mentioned technologies. Finally, the conclusions and the challenges are discussed.

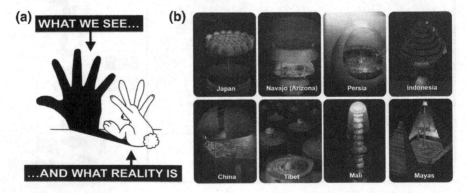

Fig. 3 **a** Schematics of the concept of modeling: models are representations of the reality (figure adapted from Mobii Arts [4]); **b** different models of the Universe by old civilizations [5]

2 Integrative Multiscale Modeling Methods

From an engineering perspective, the use of continuum modeling represents an elegant way of tracking the competition and synergies between physicochemical mechanisms. In general, it is important to develop modeling tools that can evaluate the relative impact of each mechanism on the overall efficiency and materials stability of the electrochemical device under investigation.

Continuum modeling consists of describing mathematically electrochemical, transport and thermomechanical mechanisms in the cells through a set of coupled Ordinary Differential Equations (ODEs) and Partial Differential Equations (PDEs). Depending on the aspects studied and the assumptions considered, this leads to mathematical problems of different levels of complexity to be solved numerically.

In the last decade, a new class of continuum models emerged for the modeling of devices for electrochemical energy conversion and storage. This class consists of continuum models

- which describe mathematically relevant mechanisms occurring at relevant spatial scales through within a fully continuum framework [6–9] or within hybrid continuum/discrete approaches [10–12]. These models have a hierarchical structure: that means that solution variables defined in a lower hierarchy domain have finer spatial resolution than those solved in a higher hierarchy domain. Consequently, physical and chemical quantities of smaller length-scale physics are evaluated with a finer spatial resolution to resolve the impact of the corresponding small-scale geometry. Larger scale quantities are in turn calculated with coarser spatial resolution, homogenizing the (possibly complex) smaller scale geometric features;
- which can incorporate parameters extracted from other paradigm calculations, such as Density Functional Theory (DFT) [13, 14], Molecular Dynamics (MD), or Coarse Grain Molecular Dynamics (CGMD) (Fig. 4) [3, 15, 16]. Such parameters include materials structural properties, species diffusion coefficients, and activation barriers associated to some specific electrochemical reactions.

This class of continuum models is referred here as "integrative multiscale models", and constitutes the main focus of this chapter. They aim, by construction, to significantly reduce empirical assumptions than can be done in simple multiphysics models. The reason is that they explicitly describe mechanisms in scales neglected in the simple multiphysics models.

While empirical models that are trained to experimental data provide already insights on the operation principles of fuel cells and batteries with easiness, due to their simple construction and fast computational speed, they possess several drawbacks. Specifically, empirical models are only as good as the experimental data they are trained to, and thereby do not provide the ability to extrapolate beyond the range of these data. Moreover, these models are empirical in nature, they provide little, if any, deep insight into the operation principles of the cell in relation to the materials properties.

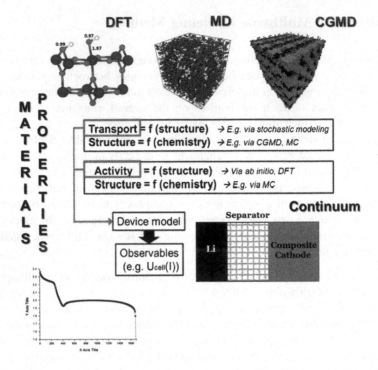

Fig. 4 A possible multiscale approach: calculating materials properties to be injected into a continuum cell model able to predict macroscopic observables

As an example, integrative multiscale modeling in the field of electrocatalysis can be developed from the use of DFT calculations [17] to estimate the values of the activation energies E_{act} of single elementary reaction kinetic steps (Table 1), and then inject them into Eyring's expressions to estimate the kinetic parameters k

$$k = \kappa \frac{k_B T}{h} \exp\left(-\frac{E_{act}}{RT}\right) \qquad (1)$$

where κ refers to the frequency prefactor, k_B and R the Boltzmann and ideal gas constants, T the absolute temperature, and h the Planck constant. More precisely, in this approach, DFT uses the so-called Nudged Elastic Band (NEB) method to find reaction pathways when both the initial and final states are known [18, 19]. NEB method consists in linearly interpolating a set of images between the known initial and final states, and then minimizing the energy of this string of images. Each "image" corresponds to a specific geometry of the atoms on their way from the initial to the final state, a snapshot along the reaction path. Thus, once the energy of this string of images has been minimized, the pathway corresponding to the minimal energy is found.

Table 1 Example of database reporting the activation energies of different elementary steps within the ORR as function of the OH coverage (from Ref. [20])

Elementary act	E_{act}^{fwd} (kJ mol^{-1})				E_{act}^{bwd} (kJ mol^{-1})				ΔE_{reac} (kJ mol^{-1})			
θ_{OH} (ML)	0	1/6	1/3	1/2	0	1/6	1/3	1/2	0	1/6	1/3	1/2
(S1) $O_{2(ads)} \rightarrow 2O_{(ads)}$	30	31	44	51	149	137	150	150	−119	−106	−106	−99
(S6) $O_{2(ads)} + H_{(ads)} \rightarrow OOH_{(ads)}$	38	35	40	35	42	56	53	56	−3	−21	−13	−21
(S2) $O_{2(ads)} + H_{(ads)} \rightarrow OH_{(ads)}$	88	90	84	80	94	100	108	115	−6	−10	−24	−34
(S3) $OH_{(ads)} + H_{(ads)} \rightarrow H_2O_{(ads)}$	19	16	15	23	80	85	94	157	−61	−69	−78	−133
(S9) $OOH_{(ads)} + H_{(ads)} \rightarrow H_2O_{2(ads)}$	24	33	23	−	46	52	57	−	−23	−18	−34	−
(S7) $OOH_{(ads)} \rightarrow OH_{(ads)} + O_{(ads)}$	0	5	4	−	158	156	199	−	−159	−152	−195	−
(S4) $2OH_{(ads)} \rightarrow H_2O_{(ads)} + O_{(ads)}$	1	5	5	−	24	25	26	−	−24	−20	−21	−

Equation (1) can be used for the calculation of the individual reaction rates at the continuum level [13],

$$v_i = k_i \prod_y a_y^v - k_{-i} \prod_{y'} a_{y'}^{v'} \tag{2}$$

where a refers to the activity of the reactants and products and v the stoichiometry coefficients. Equation (2) is in turn used for the calculation of the evolution of the surface or volume concentrations of the reaction intermediates, reactants, and products, following

$$K_n \frac{da_y}{dt} = \sum_i v_i - \sum_j v_j \tag{3}$$

where K_n is the number of reaction sites per mol of reactants.

Another example of integrative multiscale model results from the use of MD to extract diffusion coefficients [21] or of CGMD for calculation of the materials structural properties (e.g., pore size distributions) as function of the materials chemistry, which are used in turn for the estimation of the effective diffusion parameters integrated in continuum reactants transport models [22]:

$$D_{eff} = \frac{\varepsilon}{\tau} D_0 \tag{4}$$

where ε refers to the material porosity and τ to the material tortuosity. The development of modeling methods to predict the microstructural properties of components in electrochemical devices for energy conversion and storage is of paramount importance in order to understand the relationship between fabrication parameters and their performance in real applications.

Depending on the development context of the integrative multiscale models (engineer or physicist based), they would be built following top-down or bottom-up viewpoints (Fig. 5).

Top-down integrative multiscale models connect detailed macroscopic descriptions of mechanisms with global parameters representing microscopic mechanisms. On the other hand, bottom-up integrative multiscale models scale up detailed descriptions of microscopic mechanisms on global parameters to be used in macroscopic models [3].

An elegant and reliable way to develop these both types of models is through the use of the nonequilibrium thermodynamics framework [24]. This framework is supported on three postulates:

- **Postulate I**: for a system in which irreversible processes are taking place, all thermodynamic functions of state exist for each element of the system. These thermodynamic quantities for the nonequilibrium system are the same functions of the local state variables as the corresponding equilibrium thermodynamic quantities.

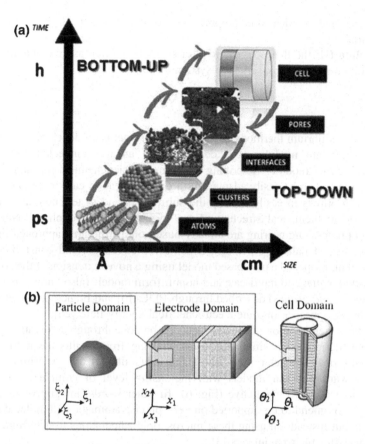

Fig. 5 **a** Bottom-up and top-down multiscale modeling approaches, and characteristic space and time scales spanned (adapted from Ref. [3]); **b** a multiscale model of a LIB (from Ref. [23])

- **Postulate II**: the thermodynamic flux densities are functions of the thermodynamic efforts, i.e.,

$$J = f(X_\alpha, X_\beta, X_\chi \ldots) \tag{5}$$

which gives at first-order linear flux/effort relationships (first-order or linear nonequilibrium thermodynamics):

$$J_i = \sum_j L_{ij} X_j \tag{6}$$

We notice that Eq. (6) can represent for instance the first Fick's law relating the molar flux of a species with its gradient of chemical potential, or any other linear combination of multiple thermodynamic efforts such as the ionic flux which can be written as the addition of a term proportional to the gradient of chemical potential and another one proportional to the gradient of the electrostatic potential.

- **Postulate III**: the thermodynamic efforts can be written as gradients of potentials (assumption of conservative systems), i.e.,

$$X_i = \nabla \varphi_i \tag{7}$$

The nonequilibrium thermodynamics framework helps on developing structured models which are modular (i.e., consisting of an interconnected network of "modules", each "module" describing a unique physicochemical mechanism) and thus reusable (it can be easily adapted from one application case to another). Such an approach allows to easily modify the submodels and to test new assumptions keeping the mathematical structure of the model and the couplings. Numerous models in process engineering are already based on a structured approach like this one using sets of balance equations, constitutive equations, and constraints [25]. Maschke et al. proposed a port-based model using a novel extension of the so-called *bond graph* language to multiscale and nonuniform models (also known as *infinite-dimensional bond graphs*) described through PDEs [6] and Franco et al. introduced this modeling approach into the electrochemical energy field [26].

Parameter identification of such models can be done through an iterative process with atomistic/molecular simulation tools, starting from multiscale models with mainly empirical character (i.e., with parameters fitted from experiments), and moving toward physical models with increasing level of complexity with the ultimate goal of being predictive (Fig. 6). In other words, integrative multiscale models do not intend to be supported on very precise atomistic and molecular level databases, but instead, on using these microscopic calculations to determine initial guess values for their parameters [3].

Due to the inherent complexity of multiscale modeling to lead multiple materials and mechanisms, one may also require hierarchical concepts to connect simulation protocols with the computing infrastructure, particularly with HPC architectures, on which the simulations can be executed (e.g., in the case of ab initio codes). Automatization of the generation of database libraries and their integration in indirect multiparadigm models are also important aspects to be considered [27]. Some software platforms allowing to create data flows, to selectively execute some computational steps, and to automatically inspect the results, are already available [28, 29]. Particularly for Lithium Ion Batteries (LIBs), an in-house flexible and scalable computational framework for integrated indirect multiparadigm modeling is reported by Elwasif et al. [30]. The framework includes routines for the codes execution coordination, computational resources management, data management, and inter-codes communication. The framework is interfaced with sensitivity analysis and optimization software to enable automatic LIB design.

For further details and discussions about the different types of multiscale models, the readers are invited to consult previous references by this chapter author [2, 3, 16, 31].

Fig. 6 Schematics of an iterative approach for the development of detailed models starting from experimental data

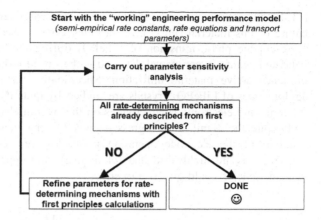

3 Application Examples

In view of the wide diversity of problems where multiscale models can be applied in energy conversion and storage devices, two illustrative cases are chosen here

- the investigation of the impact of the electrodes microstructural features on the performance and durability of the cells;
- the incorporation of detailed electrochemical models in macroscopic cell models, allowing to capture the influence of the active material chemical and structural properties on the overall cell performance.

3.1 Microstructurally Resolved Performance Models

In the context of fuel cells and batteries, several mathematical models have been developed attempting to capture the influence of the electrodes structural properties on the ionic transport properties and cell response. As these models attempt to describe transport mechanisms at different scales of porosity, for example, they can be classified as "multiscale models".

In the case of PEMFCs the so-called spherical agglomerate model with a single diameter constitutes the classical approach to describe the average electrochemical reaction rate in the electrodes. For instance, Hu et al. [32] use a classical agglomerate model with thin film of polymer and liquid water to investigate the effects of the cathode properties on the PEMFC performance. This type of approaches permit in an elegant way to capture the effects of the agglomerates radius, Pt loading and Pt particle radius, operation temperature and pressure on the PEMFC performance, and can be used to optimizing it in terms of operational conditions and electrode structure (Fig. 7) [33–35].

Another example of the application of these approaches is the work reported by Dargaville and Farrell [36] on LIBs consisting of the simulation of the discharge of a LiFePO$_4$ cathode accounting for three size scales representing the multiscale nature of this material (Fig. 8). A shrinking core is used on the smallest scale to represent the phase transition of LiFePO$_4$ during discharge. The model is then validated against experimental data and is then used to investigate parameters that influence active material utilization. Specifically, the size and composition of agglomerates of LiFePO$_4$ crystals are studied by quantifying the relative effects of the ionic and electronic conductivities on the overall electrode capacity.

Despite the significant progress achieved using these models on the understanding of the electrode operation principles, real composite electrodes and agglomerates are highly irregular (cf. Fig. 2), and approximating them by independent spheres could give rise to errors.

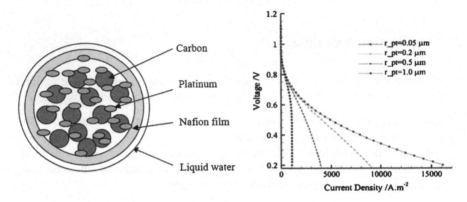

Fig. 7 An example of agglomerate model and calculated cell performance curve for different agglomerate diameters (from Ref. [32])

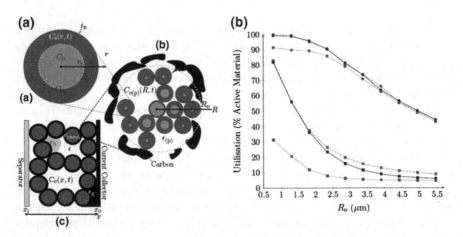

Fig. 8 a Schematic of the three size scales in the LIB cathode model of Dargaville and Farrell (crystal, particle, electrode); **b** particle scale utilization versus agglomerate size for different discharge rates (from Ref. [36])

Fig. 9 Distribution of the normalized oxygen concentrations in a PEMFC cathode electrode under an overpotential of 0.65 V (**a**) in comparison with that under an overpotential of 0.85 V (**b**). The value of the normalized concentration changes from 1 (*red*) to 0.05 (*blue*) (from Ref. [37]) (Color figure online)

Zhang et al. [37] reported an investigation consisting of acquiring three-dimensional microstructure of a PEMFC cathode using FIB/SEM tomography (Fig. 9). Oxygen diffusion and the associated electrochemical reaction in the microstructure were simulated using explicit pore-scale modeling. The simulations were then compared with the results predicted by a spherical agglomerate model using an average diameter estimated from the three-dimensional microstructure. The authors found that the spherical agglomerate model substantially overestimated the reaction rates and the overpotential. This implicates that the spherical agglomerate model needs to be used with care in the PEMFC electrode design as its diameter is frequently a fitting parameter rather than a geometrical description of the agglomerates.

In the context of LIBs, Roberts et al. [38] recently proposed a model which leads with the anisotropic, lithiation-induced mechanical deformation of cathode materials. The model permits simulations directly resolving the cathode particles and the surrounding electrolyte through the use of the finite element method applied to FIB/SEM reconstructions and artificial particle size distributions (Fig. 10). The model accounts for a Butler–Volmer-based approach of the electrochemistry coupled on the fly with a mechanical model involving a quasi-static elastic constitutive equation with anisotropic swelling strains that are assumed to be linear in the local lithium concentration. From extensive simulations, they find that for particles that swell isotropically, the largest stresses are located at the particle-to-particle contacts, due to neighboring particles constraining the swelling of a particle.

Furthermore, there are very few efforts so far aiming to predict the structural properties of composite electrodes as a function of the fabrication parameters and

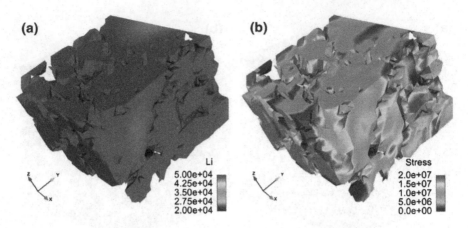

Fig. 10 Calculated snapshot showing that high stress concentrations arise near particle contacts. **a** Lithium concentration in the solid particles, in units of mol/m^3. **b** Stress in the solid particles, in units of Pascal

the composition. For instance, stochastic or Monte Carlo (MC) models consisting of multiparticles energy-based spatial arrangement optimization are just emerging and used to mimic electrodes fabrication and to estimate relevant structural and transport properties [39, 40]. Because of their fully empirical parameterization, these techniques cannot predict the influence of the chemical composition (e.g., solvent and active material chemistries) on the arising electrode self-organization and structural properties (Fig. 11).

Many publications have been reported particle-resolved simulations to predict conductivity within porous composite electrodes, in particular in relation to Solid Oxide Fuel Cells (SOFCs) [41–46]. In this case, reported methods consist of packing randomly spherical particles into a cubical region in order to generate in silico composite structures constituted by both ion-conducting and electron-conducting particles [41]. The particle network is discretized using a meshing that can resolve the particles themselves and their intersections. Charge conservation equations are solved to predict current through the network (Fig. 12). According to Kee et al. predicted effective conductivities with these methods are significantly smaller than those predicted with conventional percolation theory, which highlights the importance of the three-dimensional interparticle connectivity in determining the overall cell performance [34].

Since the 1990s, LIBs are the most widely used energy storage devices for portable applications. However, commercially available liquid and organic polymer electrolytes induce poor thermal stability and low resistance to leakage and safety risks due to the presence of volatile organic solvents. Additionally, such conventional electrolytes cause decomposition through electrode/electrolyte parasitic reactions due to their narrow voltage window. Upon cycling a risk of electrolyte consumption exists and may lead to cell failure. Therefore, such batteries may suffer from severe technological issues preventing them from being a satisfying answer to the rising

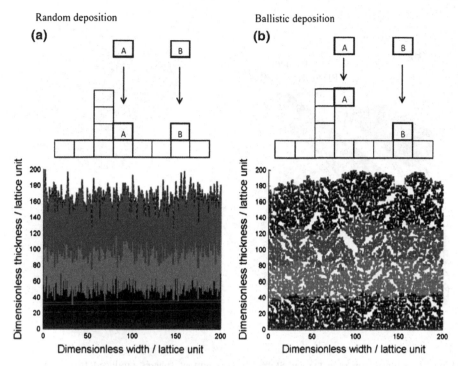

Fig. 11 The layer formed during the deposition of an active material. **a** Random deposition, **b** correlated growth (ballistic) deposition. The different *colors* correspond to different loadings (deposition of 1000 particles) (from Ref. [39]) (Color figure online)

needs of new emerging applications (electric/hybrid vehicles, sustainable large-scale energy storage), or even more specific fields with strong added value such as portable microelectronic devices and microelectromechanical systems (MEMS) [47]. At present, the major concerns for improving the current technology including safety, reliability, and lifetime may be concentrated into a common solution relying on all-ceramic (or all-solid-state) LIBs (ASLIBs). ASLIBs potentially exhibit the required gravimetric power and energy density for practical uses in the new application fields of electromobility and grid storage due to their advantages with respect to LIBs batteries (easy to miniaturize, long shelf lives, stability of performance with temperature, etc.). However, while they can provide advantages over liquid electrolytes in terms of safety, reliability, and simplicity of design, ASLIBs face interface problems with the electrodes, their ionic conductivity being generally lower than those of liquids and polymers [48]. The choice of the most suitable solid electrolyte for a particular application will depend on several factors, such as operating parameters (e.g., voltage range, temperature) and battery design (e.g., rigid or flexible). Ideally speaking, the electrodes in ASLIBs have to show good electrochemical properties with low kinetic limitations. Furthermore, they should provide good mechanical stability with small volume changes upon cycling so as to limit internal

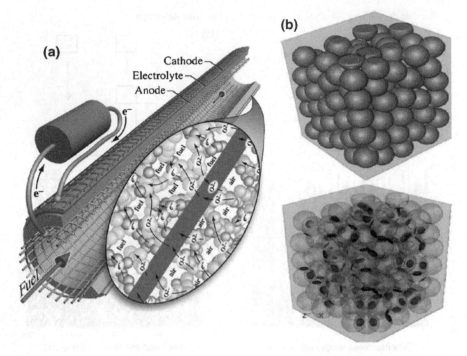

Fig. 12 a Illustration of a tubular SOFC; **b** a system of spheres produced from a packing algorithm. The lower image highlights the contact areas between the spheres after artificial sintering (from Ref. [41])

stress within the electrode and then crack formation. In particular, the electronic conductive additive should be easy to disperse to favor the sintering process and overcome the well-known limitations of carbon materials. In this context, the formulation and the fabrication process of the composite electrode have to be rationally designed to ensure an optimal three-dimensional ionic and electronic percolation, while minimizing the amounts of electrolyte/electronic conductor, both acting as inert materials. The optimization of the composite electrodes formulation is the only way for reaching high energy density batteries (Fig. 13).

The so-called Discrete Element Method (DEM) is a well-suited technique to address the particles rearrangement/deformations upon mechanical compression in granular materials in ASLIBs [50]. It explicitly accounts for the mechanical interactions between the individual particles, and allows capturing particles deformation and cracking. The method has been already used for the simulation of morphological changes of composite materials in multiple applications, including SOFC ceramic electrodes [41, 51] and some preliminary adaptation has been carried out recently by Franco et al. to simulate the micro/mesostructural properties of ASLIB composite electrodes (Fig. 14) [52]. The method resolves Newton's equations for the trajectory of individual particles and/or aggregates from their mechanical properties (Young modulus) and interaction mathematical laws

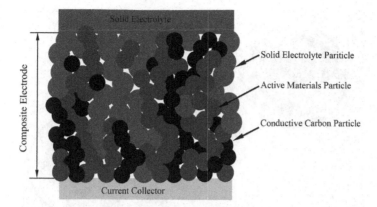

Fig. 13 Illustration of the structural complexity of an ASLIB composite electrode [49]

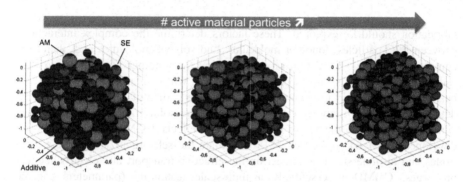

Fig. 14 Example of predicted electrode structure as a function of the number of active material particles per unit of volume. Results obtained with the DEM model being developed by Franco et al. for the prediction of the structural properties of composite solid electrodes (generic materials; *AM* active material particle, *SE* solid electrolyte particle) (from Ref. [52])

accounting for stress-driven deformation (simulated as an overlap between particles). The application of pressure on the powder can be carried out by decreasing the size of the simulation box to reach the experimentally observed electrode density. This can lead to the prediction of the dissipative dynamics of the multiparticle/aggregate system and of the final mesostructure. Mechanical stress testing can be performed for evaluating the level of accuracy of the model for predicting micro/mesostructural changes under mechanical stresses. The use of percolation theory and Fast Fourier Transform methods can be used for the determination of the effective conductivities of the calculated micro/mesostructures.

Electrode microstructures can also be generated in silico by atomistic methods. To improve the understanding of the electrode structure, for example, in PEMFCs, the effects of applicable solvent, particle sizes of primary carbon powders, wetting properties of carbon materials, and composition of the ink used during the

Fig. 15 A typical CGMD-calculated PEMFC electrode structure (from Ref. [54])

fabrication should be explored. These factors determine the complex interactions between Pt/C particles, ionomer molecules, and solvent molecules and, therefore, control the electrode formation process. Mixing the ionomer with dispersed Pt/C catalysts in the ink suspension prior to deposition will increase the interfacial area between ionomer and Pt/C nanoparticles [53]. The choice of a dispersion medium determines whether ionomer is to be found in the solubilized, colloidal, or precipitated forms. With the aim of predicting these effects, CGMD models have been developed by Malek et al. [54, 55] to simulate the self-organization of the electrodes and to understand its impact on the effective transport and electrochemical properties. CGMD is essentially a multiscale technique (parameters can be extracted from DFT and/or MD calculations). CGMD simulations have been employed for characterizing microstructure of the electrode in view of effect of solvent, ionomer, and Pt particles (Fig. 15).

Furthermore, in the last decade, Franco et al. have proposed multiscale modeling approaches which are able to capture the interplay between the electrode and materials microstructure evolution and the local cell operation conditions [56–62]. The particularly novel character of the authors' cell models is the simulation of the feedback between detailed electrochemistry and transport with materials aging mechanisms (Fig. 16): that means at each numerical simulation time step, the model describes how the calculated local conditions impact local materials degradation kinetics, simultaneously to how the materials degradation affects, in the next time step, the local conditions. Then, the models can predict cell durability (i.e., potential evolution at fixed operation conditions, in contrast to the majority of the models available in the literature where this is impossible as they fix the potential as an input parameter) [63].

Malek and Franco carried out CGMD simulations to build a structural database for PEMFC cathode electrodes with different C contents in terms of interpolated mathematical functions describing the impact of the C mass loss (induced by corrosion) on the evolution of the ionomer coverage on Pt and C, the electronic

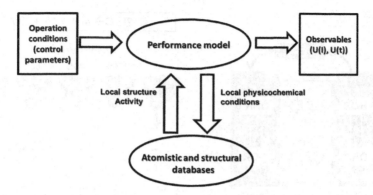

Fig. 16 Multiscale modeling approach by Franco et al. for the prediction of PEMFCs durability

conductivity of the C, the C surface area, and the Pt surface area (which reorganizes during the C corrosion process) [22]. These functions are then integrated into a performance model to simulate the impact of C corrosion on the cell performance decay following the algorithm described in Fig. 16. Such a performance model incorporates detailed description of the electrodes electrochemistry (Hydrogen Oxidation Reaction/ORR in the anode, ORR and Carbon Oxidation Reaction in the cathode) and ionic and reactants transport. The resulting performance models are able to describe the morphological changes of the catalyst layer/membrane (evolution of the porosity/tortuosity calculated by CGMD) induced by the electrochemical/chemical degradation, and simultaneously, to describe how these morphological changes affect the effective transport properties (water, proton, oxygen) and overall cell performance. The influence of several operation scenarios on the prediction of the cell potential decay can then be investigated (Fig. 17). This model has been also extended recently to predict the PEMFC durability due to the membrane degradation [64].

A recent work prepares the extension of this work by addressing the question of the PSD water filling dynamics versus the active area variation, in the context of open-cathode PEMFCs [65]. The authors propose a dynamic multiscale model describing two-phase water transport, electrochemistry, and thermal management within a framework that combines a Computational Fluid Dynamics (CFD) approach with a microstructurally resolved model predicting the water-filling dynamics of the electrode pores and the impact of these dynamics on the evolution of the electrochemically active surface area (ECSA). The model allows relating for the first time the cathode electrode structure to the cell voltage transient behavior during experimental changes in fuel cell temperature. The effect of evaporation rates, desorption rates, and temperature changes on the performance of four different electrode pore size distributions are explored using steady state and transient numerical simulations. The model helps to understand experimentally observed thermal and electrochemical system dynamics, which is essential for the development of proper control strategies. It has been shown that for the relatively

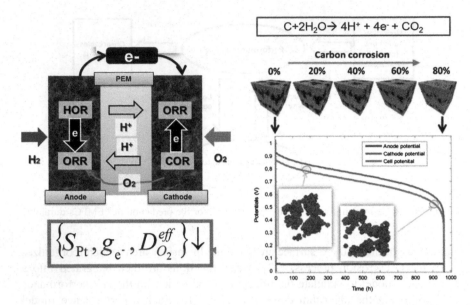

Fig. 17 Schematics of the PEMFC cathode carbon corrosion principles and simulation result from the integration of the CGMD database into a performance model (figure adapted from Ref. [22])

dry open-cathode system the evaporation rate and the liquid water sorption constants are crucial for proper representation of the cathode electrode performance. The dynamics of the voltage response with respect to an increase in cell temperature are dominated by water desorption dynamics of the Nafion® thin film in secondary pores. Of course, these aspects would also add interesting interplays with the carbon corrosion dynamics, and with other materials degradation mechanisms as well, in an operating PEMFC.

Tracking the influence of the detailed electrode microstructure on the cell performance starts to be recognized as a fundamental problem in several emerging technologies, such as the lithium air batteries (LABs) [66]. Within this sense, a pioneering multiscale model of a LAB considering the cathode pore size distribution was proposed by Franco et al., where the morphology of the discharge product, Li_2O_2, is assumed to be thin films covering the surface of the pores [67]. In the model, the active surface area degrades during discharge because of three reasons:

- first, the effective radius of pores decreases due to Li_2O_2 coverage;
- second, small pores may be fully choked;
- third, thick Li_2O_2 film may block the electron tunneling process, rendering the surface inactive.

Simulation results reveal that the end of discharge in cells with cathode electrodes made of Super P and Ketjen Black carbons is caused by unavailable surface area near the air inlet, rather than the full choking of pores. Larger discharge

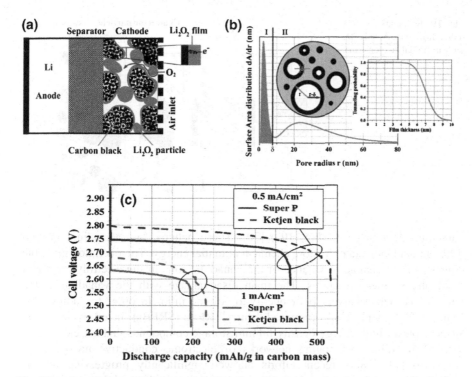

Fig. 18 **a** Schematics of Franco et al. LAB cell model; **b** example of surface area distribution (associated to a pore size distribution) and tunneling function implemented in the model; **c** examples of simulation results for two types of carbon structures and for two values of discharge current densities (figure adapted from Ref. [67])

capacity is found in the Ketjen Black cell because its high specific surface area leads to slower Li_2O_2 thickness growth rate: this is an experimentally known fact which has been reproduced for the first time with this model. The model demonstrated the compromise between high surface area (small pore size) and large pore radius (slow degradation rate of active surface) (Fig. 18). These calculations conclusions have been later reproduced by other authors [68, 69].

Other emerging battery technologies provide new interesting challenges in the understanding of the electrode structure/performance relationships still unsolved, as described in the next paragraphs.

Batteries for large-scale power grid storage require durability for large numbers of charge/discharge cycles as well as calendar life, high efficiency, an ability to respond rapidly to changes in load or input, and reasonable costs. Redox flow batteries (RFBs) promise to meet many of these requirements. In these batteries, the electrochemically active species are dissolved in an electrolyte and stored externally, the main advantage is that they can offer almost unlimited capacity simply using larger and larger storage tanks. These batteries were originally developed by NASA in the 1970s, but popularized in the late 1980s wth the development of the

Fig. 19 Schematics of slurry redox flow battery (SRFB) with colloidal suspensions

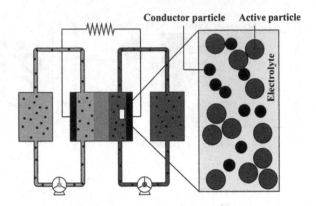

vanadium flow batteries (Fig. 19a) [70]. These batteries use vanadium ions in different oxidation states to store chemical potential energy. The operation principle consists of the change of valence of the vanadium ions in the electrolyte without solid phase battery reactions. The main disadvantage with the vanadium redox batteries is their relatively poor energy-to-volume ratio. In order to address this issue, in 2011 MIT (USA) proposed "slurry" RFBs (SRFBs) fueled by semisolid suspensions of high-energy-density lithium storage compounds that are electrically "wired" by dilute percolating networks of electronic conductive nanoparticles (Fig. 19) [71]. More recent efforts allowed significantly progressing on the understanding of their electrochemical characterization with multiple materials [72, 73]. Despite the progresses achieved, there is still a severe lack of understanding of the SRFB operation principles in particular in relation with the influence of the slurry structural properties on their overall electrochemical response.

SRFBs hybridize aspects of static and flowing batteries and thereby they provide improved energy density and reduced cost. However, a viscous, conductive suspension may incur efficiency loss mechanisms not encountered in conventional flow batteries, such as the viscous energy dissipation and the extension of the electroactive region outside the cell stack. SRFBs operation principles arise from a strong coupling between electrochemical mechanisms and fluid mechanics. Understanding the behavior of the fluids with particles suspensions is of crucial importance toward the optimization of SRFBs as their performance significantly depends on the facility of the electronic conductive particles to dynamically percolate. For instance, a deep investigation of the effect of flow rates, formulations, suspension particles sizes, impact of carbon additives, electrolyte purity, etc. on the battery performance is needed. Furthermore, the use of turbulent regimes for the mixing of the suspension particles remains unexplored. The choice of monophasic or biphasic active materials (versus Lithium insertion) may also impacts the cyclability of these batteries.

Out of the context of batteries, there have been significant progresses in understanding the rheological properties of simple dynamical suspensions, e.g., very dilute and semidilute suspensions, however, understanding the flow of more

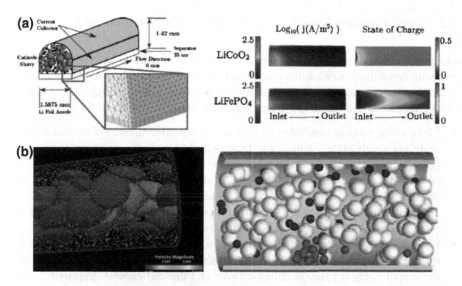

Fig. 20 **a** Continuum model of a SRFB [74, 75]; **b** DPD modeling of red blood cells [76]

complex suspensions, e.g., dense suspensions, random-shaped particles, suspensions composed of particles that interact remains a great challenge. One possible approach is to apply standard continuum CFD methods. However, this reveals to be a very complex task as it involves significant effort in tracking boundaries between different fluid and solid phases, usually involving various meshing, moving grids, and interpolation schemes to account for motion of the rigid bodies.

Only very few theoretical studies have been attempted so far on SRFBs [74, 75]. All the models available in the literature consider the slurry as a continuum. The instantaneous structure arising from the particles percolations is assumed to be static and it is used for solving performance models through effective (Li^+ and electronic) conductivity properties calculated from averaged tortuosities and porosities (Fig. 20a). Furthermore, there is a tremendous lack of models providing insights about the detailed behavior of the suspension particles in dynamical conditions.

As pointed out recently by Franco et al. [77], the Dissipative Particle Dynamics (DPD) method can be interesting to predict the dynamic percolating of the different suspensions particles (active material and carbon additive particles) as function of the slurry composition and operation scenarios (e.g., constant or dynamically changing global flow rates, channel geometry). The DPD method consists of explicitly resolving the interaction between particles in colloidal static or dynamic fluids. It can be applied to study the rheology of dense supensions of spheres, rods, and disks. This technique, originally developed in the early 1990s, [78] explicitly resolve the particles' trajectories by considering the interparticles shocks and associated energy dissipation. For instance, the DPD approach is currently used for the simulation of blood cells (Fig. 20b).

3.2 Performance Models with Detailed Electrochemistry

Many atomistic and continuum models available in the literature have demonstrated powerful capabilities to track the electrochemical cell operation principles, but they still present several drawbacks that make them not appropriate enough to predict the cell performance and durability under realistic operation conditions [63]:

- Many of the models available in the literature consider kinetic submodels which represent steady-state regimes with time-independent local operating conditions [79]. Indeed, they do not take into account the impact of the modeled aging phenomena on the variation of the structural and physicochemical properties of the cell materials (e.g., impact of the aging mechanisms on the variation of charge conductivity, variation of porosity, variation of contact resistances, etc.), [80] in contrast to the example discussed earlier in Sect. 3.1 (cf. Fig. 17).
- Potentiostatic–potentiodynamic simulations. In most of the available kinetic degradation models, the Butler–Volmer electrode potential is the input variable, the output being a material corrosion rate and the cell current. Implicitly, it is assumed that the potential of the materials is equal to the external/macroscopic applied potential.
- Use of the classical Butler–Volmer theory. This empirical theory, largely used in the field of electrochemical devices age describes electrochemical (electron transfer) reactions on ideal planar electrodes. Using empirical Butler–Volmer equations written in terms of the electrode potential [81–83], oxide formation and corrosion reactions are implicitly supposed to take place in the bulk, just outside the interfacial electrochemical double layer—EDL—(i.e., far from the electrified substrates). Thus the possible interplaying of these aging mechanisms with the EDL structure, expected in realistic environments, is not taken into account [63].

Resolving the detailed EDL structure moving beyond the classical equilibrium theories (Fig. 21) and its interplay with REDOX reactions it is crucial in order to help on answering questions such as:

- How different is the ionic transport at the bulk, surface, and interface scales?
- What is the structure of the EDL at the active particle/electrolyte interface, and how does this EDL affect the cell polarization and transient behavior?
- What is the relative impact of these mechanisms on the overall cell performance?

A few attempts have been reported in the literature so far for describing EDLs in fuel cells and in electrode systems where the ionic transport is usually modeled within the Poisson–Nernst–Planck (PNP) approach under the diluted solution approximation but without taking into the account explicitly the role that solvent may have on the effective transport properties of the media [84]. Furthermore, numerous continuum theories have been reported since 80 years to describe multicomponent liquid mixtures at the bulk level. These models have been used and

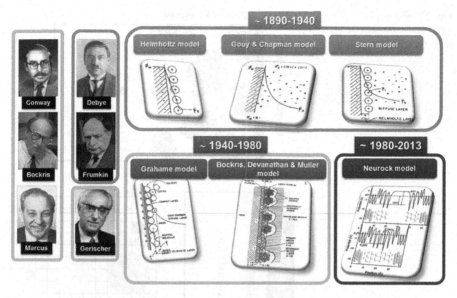

Fig. 21 Some major contributors and associated periods of theories describing the EDL structure. All these reported theories refer to the EDL at equilibrium conditions, thus conditions not fully representative of the ones found in operating electrochemical energy devices

further developed for calculating the static dielectric constant of mixed-solvent electrolyte solutions. There is, however, a lack of understanding on the structural properties of the liquid mixtures at the vicinity of charged surfaces.

Franco et al. [85, 7] develop since 14 years electrode/electrolyte interface (EDL) continuum models supported on a statistical mechanics approach. Their theory was recently further developed to capture the impact on the overall electrode behavior of the electrolyte composition in terms of solvent, charged polymers, and ions concentration, for a large diversity of cases, from diluted solutions to ionic liquids [86]. From its continuum character, the theory is particularly useful for the simulation of interfacial electrochemical mechanisms within multiscale frameworks scaling up atomistic and molecular level properties on overall performance cell models.

Franco et al. EDL model consists of two distinct regions (Fig. 22): an external layer (EL) and an inner layer (IL). The EL model describes the ions and counterions transport, affected by the solvent composition and/or the morphology of a charged polymer eventually present in the solution. The EL model accounts for both the finite size effect of the species and the relative size effect between them. The IL model describes the coverage evolution of the intermediate reaction species (involved in possible REDOX reactions), adsorbing/desorbing solvent molecules and ions on the substrate surface.

The charge density σ of the electrode (active particle) in Fig. 22 is related to the imposed electronic current density J through

Fig. 22 Model
implementation scheme of the
EDL region (from Ref. [86])

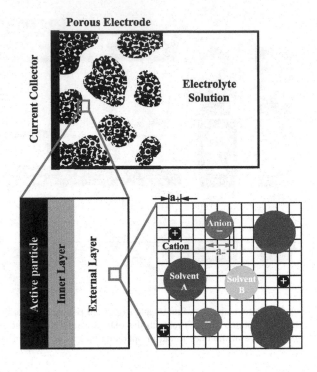

$$J - J_{FAR} = -\frac{\partial \sigma}{\partial t} \tag{8}$$

where J_{Far} is the faradaic current density which is proportional to the algebraic addition of the kinetic rates v_i associated to electrochemical reactions happening on the surface according to

$$J_{FAR} = F \sum_i (\pm n_i) v_i \tag{9}$$

where F is the Faraday constant and n_i is the stoichiometric coefficient. The kinetic rates are given by Eq. (2) where the kinetic parameters are given by the modified Eyring's expression from the Transition State Theory (thus, similar to Eq. (1))

$$k_l = \kappa \frac{k_B T}{h} \exp\left(\frac{-E_{act,l} + f(\sigma)}{RT}\right) \tag{10}$$

where $E_{act,l}$ is the activation energy without influence of the interfacial electric field (it may be estimated from DFT/NEB calculations), κ is the frequency prefactor (it can include organizational partition functions), k_B and R are the Boltzmann and the ideal gas constants, respectively, T is the absolute temperature, and h is the Planck constant. $f(\sigma)$ is a function of the charge density σ, the so-called surface potential different of zero only for electrochemical steps. $f(\sigma)$ is also proportional to the electrostatic potential drop through the IL between the electrode surface and the electrolyte. We note that the exponential argument, $E_{act,i} + f(\sigma)$, is an effective activation energy. The authors demonstrated that $f(\sigma)$ in Eq. (10) is function of the surface concentration of adsorbed ions on the electrode c_j and Γ being the dipolar contribution to the total electrostatic potential jump through the IL:

$$f(\sigma) = \frac{F}{N_A \varepsilon_0} \left(\sigma + F \sum_j Z_j c_j - \Gamma \right) H \tag{11}$$

Γ is a function of the coverage of intermediate reaction species and of solvent adsorbed molecules quantities calculated through appropriate mass balance equations [86].

The equations describing the charge distribution in the EL are derived from the minimization of the free energy functional $F = U - TS$ written in terms of the electrostatic potential and the mobile particles concentration in the electrolyte (ions, solvents, polymers). The entropic term (TS) relates to the total number of spatial configurations that the mobile particles can adopt over the available sites

$$TS = k_B T \ln(W) \tag{12}$$

where W is the number of combinations for the distribution of i types of particles in a regular grid (cf. Fig. 22). The internal energy (U) can be written in general

$$U = \sum_{ij} a_{ij} c_i c_j + F \sum_i z_i c_i \varphi - N_A \sum_i c_i \vec{p}_i \cdot \vec{\nabla}\varphi \tag{13}$$

where a_{ij} is the interaction energy between particles of type i and j (parameter which can be related to the solvation energies of an ion for instance), φ is the electrostatic potential, and $\vec{p}_i$ is the effective dipolar moment of the particle of type i in the EL.

By taking the derivative of the free energy on the composition, which defines the chemical potential, Franco et al. derived a set of PDEs to describe ionic transport within the EL, as

$$\frac{\partial c_i}{\partial t} = B_i RT \, \nabla^2 c_i + B_i \vec{\nabla} \cdot \left(c_i \vec{\nabla} \sum_i a_{ij} c_j \right) + B_i z_i F \vec{\nabla} \cdot \left(c_i \vec{\nabla}\varphi \right)$$
$$- N_A B_i \vec{\nabla} \cdot \left(c_i \vec{\nabla}\left(\vec{p}_i \cdot \vec{\nabla}\varphi \right) \right) + f_1(\beta_i, c_i) - f_2(\beta_i, c_i) \tag{14}$$

where $D_i/RT = B_i$ (i.e., function of diffusion coefficients which can be calculated in principle from MD) and f_1 and f_2 are nonlinear functions of the concentrations accounting for the finite and relative size effects given in Ref. [86]. The first term on the right side of Eq. (14) represents the contribution from the configurational entropy term; the second one accounts for the interaction energy among the species and is one of the extensions of the NP approach provided by the present work; the third one represents the electromigration; meanwhile, the fourth one accounts for the dipoles interaction with the electric field and represents another extension of the NP approach arising from this work. In order to separate the contribution of the ions and the polar solvent in the charge term of Poisson's equation, the authors split it into free charge (displacement) and polarization charge

$$\varepsilon_0 \nabla \cdot \vec{E} = -\varepsilon_0 \nabla^2 \varphi = F \sum_i z_i c_i - N_A \sum_s \left[\left(\vec{\nabla} \cdot \vec{p}_s \right) c_s + \vec{p}_s \cdot \vec{\nabla} c_s \right] \qquad (15)$$

We notice that, in contrast to the classical Poisson's equation, Eq. (15) is independent on the average electric permittivity of the electrolyte $\varepsilon = \varepsilon_r \varepsilon_0$. Note that here $\vec{p}_s$ refers to the effective orientational dipolar moment of the solvent molecule or, in the case of ionic liquids, the dipolar moment of the ion. The orientational dipolar moment is a function of the electric field $-\nabla \varphi$.

The model presented above predicts heterogeneous profiles of electric permittivity, at the vicinity of an electrode as function of the electric field for different solvents compositions representative of LIB electrolytes (Fig. 23). We notice that the electric permittivity significantly deviates from its bulk value with the electric field magnitude.

The model has been used to investigate the influence of the solvent composition on the ions and counterions concentration profiles within the EL. Simulations have been carried out for electrolytes with different solvent compositions: EC, EMC, and DMC (Fig. 24) and the results arising with this theory have been compared to the ones obtained with the classical PNP theory.

For the LIB case, regardless of the solvent used, Fig. 24b, c shows drastic differences in outer layer simulation results between PNP and the Franco et al. theory. First, PNP gives much higher Li$^+$ concentration near the IL than the theory. This can be explained by the overestimated electrolyte permittivity near the IL for

Fig. 23 Average electric permittivity calculated for different solvents composition (from Ref. [86])

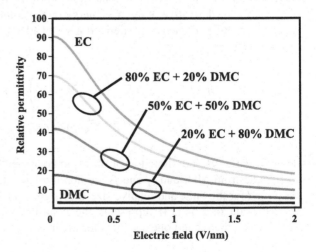

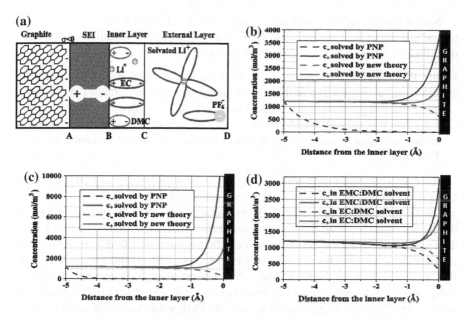

Fig. 24 **a** Simulated interface. **b** Simulated steady-state ion concentration profiles across the EL. The solvent is EC:DMC in mass ratio 4:1. **c** Simulated steady-state ion concentration profiles across the EL. The solvent is EMC:DMC in mass ratio 4:1. **d** Simulated steady-state ion concentration profiles across the EL. The solvent is either EC:DMC or EMC:DMC both in mass ratio 4 (from Ref. [86])

the PNP approach, where a constant solvent permittivity is used, regardless of the external electric field. In our theory, however, close to the IL a strong electric field is present, which lowers the electrolyte permittivity significantly. To maintain the same steady state, Franco et al. theory requires much less net charge (c_+ to c_-) close to the IL.

Second, the anion concentration in our theory is smoothly varying from the bulk value, near the bulk solution, to a lower value, near the IL. Yet, PNP gives a discontinuous spatial derivative of anion concentration at the junction between outer layer and bulk solution, which seems artificial due to the choice of diffuse layer thickness as well as the boundary condition. Hence, Franco et al. theory has better adaptability to the boundary condition of constant salt concentration in the bulk solution. Figure 24d shows that within the same approach, i.e., Franco et al. theory, the net charge near the IL will increase if there is an absence of highly polar solvents in the electrolyte solution. Indeed, while EMC and DMC both possess small dipole moments, the inclusion of EC may greatly reduce the maximum lithium concentration in the EDL, therefore lowering the risk of salt precipitation. This phenomenon can be attributed to the increasing of the electric field for the EMC:DMC case, close to the IL which produce a higher net charge in this region.

Notice that it could be possible to couple on-the-fly within an integrative multiscale algorithm this model with the atomistic theory of solid/liquid interfaces described in Chap. 1 of this book. More precisely,

- the EDL model described here can provide to the PCM model of Chap. 1 a calculated average value of the electric permittivity as function of the simulated conditions (e.g., electrolyte composition, type of solvent);
- Based on this value, the PCM model would be able to calculate electrochemical activation barriers which will then be incorporated as numerical databases in the EDL model.

Another important aspect still challenging to be incorporated in EDL models in energy conversion and storage is the presence of polymers and charged polymers which are typically present since the fabrication process (like the binder) or formed during the operation (like the so-called Solid Electrolyte Interphase (SEI) in LIBs). The structural features of such polymers can be captured by CGMD simulations, as recently demonstrated for the case of PEMFCs [87–89]: simulation results show that the hydrophilicity degree of the substrate can strongly impact the interfacial morphology of Nafion® thin films and thus the protonic charge distribution over space (Fig. 25). This is expected to impact the EDL structure which will impact in turn the effectiveness of the ORR.

The ionomer film structure will be impacted by the catalyst/carbon oxidation state (which determines its hydrophilicity). As the distribution of charge at the vicinity of the substrate is strongly affected by the ionomer structure, the surface hydrophilicity is expected to impact the proton concentration at the reaction plane, and nonuniform reaction rates are expected inside the CL. It is important to note that the hydrophilicity of the Pt is expected to evolve during the PEMFC operation as its oxidation state changes (it becomes more oxidized when the ORR occurs at its surface). Thus, the structure of Nafion at the interface is also expected to evolve upon the PEMFC operation. All these structural features are expected to strongly

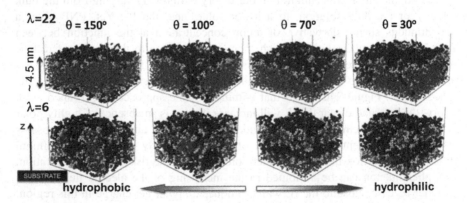

Fig. 25 Calculated impact of a substrate hydrophilicity on the structural features of an ionomer/water system, for two hydratation levels (adapted from Ref. [88])

impact the ORR kinetics through the polymer poisoning of the catalyst and the effective ionic transport and water uptake properties of the thin film.

Besides the possibilities of developing multiscale cell models incorporating mean field elementary kinetics, [7, 8, 13] the Kinetic Monte Carlo (KMC) approach is highly relevant for investigating electrochemical reactions in devices for electrochemical energy conversion and storage which show sensitivity to the structure of the active materials [21]. For instance, in the context of LIBs, Van der Ven and Ceder [90] used KMC simulations to study lithium diffusion through a divacancy mechanism in layered Li_xCoO_2. In the frame of PEMFCs, the description of the relationship between the catalyst/electrolyte interfacial charge distribution and its evolution represents a significant issue for the understanding of the REDOX kinetic processes and the prediction of the associated effective catalyst activity. Atomistic methods such as the DFT approach constitute powerful tools to study isolated redox events but usually lack on treating competing events and involved species in complex reactions where both the electric field and the temperature are not zero. In this sense, in order to describe the electrode surface coverage dynamics in relation to processes like adsorption, desorption, and reactions, the implementation of a KMC method arises as a highly relevant technique to get a proper description of surface electrocatalytic events. Within this sense, Zhdanov [91] implemented the KMC approach to study the O_2 reduction on Pt(100) finding that the impact in the overall kinetics of the lateral interaction between adsorbates is not so significant as the O_2 adsorption rate. Still Zhdanov and coworkers used the KMC approach to analyze the impact of the electric field fluctuations in electrochemical reaction kinetic rates within the EDL region [92].

Very recently, Quiroga and Franco, developed an innovative multiparadigm model of a PEMFC cathode which couples an atomistically resolved model of surface electrochemical reactions with continuum models describing transport of charges and reactants in the electrolyte [11]. The atomistically resolved model is based on a new KMC algorithm developed by the authors and which extends the so-called Variable Step Size Method (VSSM) in order to account explicitly for the electrochemical conditions (Fig. 26) describing through a lattice approach, species surface diffusion, and reactions between adspecies and adsorption/desorption of the reactants/products.

This multiparadigm model arises in particular thanks to the direct (on the fly) coupling of the KMC algorithm describing the reactions with the nonequilibrium electrolyte/active material interface EDL model described earlier in this chapter and reported in detail in Ref. [86]. The continuum models are supported on a set of coupled differential equations solved in the preexisting MS LIBER-T (_Multiscale Simulator of Lithium Ion Batteries and Electrochemical Reactor Technologies_) simulation package developed by Franco (Figs. 27 and 28) [3, 93]. MS LIBER-T is coded on an independent C/Python language basis, highly flexible and modular as based on the infinite-dimensional bond graph formalism (see Sect. 2).

The overall resulting algorithm (coupling KMC with the continuum framework within MS LIBER-T) allows us to investigate how the macroscopic electrode performance responds to nanoscopic and atomistic scale mechanisms by capturing

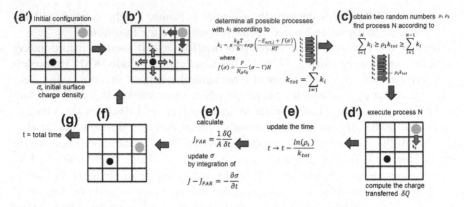

Fig. 26 The electrochemical VSSM developed by Quiroga and Franco [11]

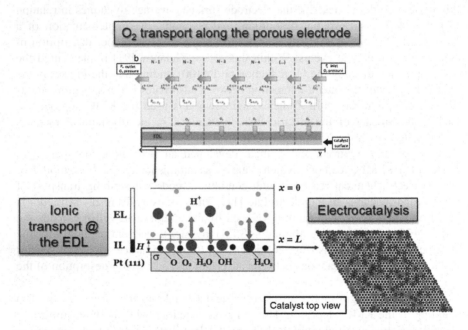

Fig. 27 Schematics of the multiparadigm model by Quiroga and Franco of a PEMFC cathode, where the transport is solved through the porous electrode thickness and through the EDL thickness, and where the detailed electrochemistry is solved with 2D resolution on the catalyst surface (adapted from Ref. [11])

details such as the surface morphology, inactive sites distribution, nanoparticle facets and sizes. This phenomena cannot be addressed with state-of-the-art continuum kinetic models supported only on the mean field approximation which neglects, by construction, adspecies surface diffusion phenomena, for example.

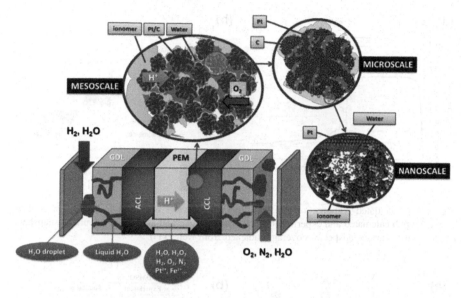

Fig. 28 Schematics of the scales considered in MS LIBER-T for the simulation of the PEMFCs

As an application example, a Pt(111)-based PEMFC cathode was studied. The model allows predicting simultaneously electrochemical observables (potential vs. time, polarization curves, etc.) and the associated surface intermediates reaction coverage (Fig. 29).

The model by Quiroga and Franco is readily applicable for other reactions where surface heterogeneity plays an important role. Indeed, we believe that the approach can offer interesting capabilities for the investigation of the ORR mechanism in LABs or the polysulfide cascade reactions in lithium sulfur battery cathodes.

Indeed, an extension of the model in Ref. [67] has been recently proposed accounting for both formation of Li_2O_2 in solution phase (solution phase reaction) and formation of Li_2O_2 as thin film (surface-limited reaction) (Fig. 30a). The extent to which the reaction is in solution phase mode is described through an escape function, which quantifies the O_2^- radicals to reach the largest open space of the cathode where they may disproportionate to form Li_2O_2 particles. Simulations and comparison with experimental data arising from two cells with TEGDME and DMSO as solvents, respectively, confirm that the escape rate is higher in a high donor number solvent such as DMSO (Fig. 30b). Further details for the ORR are under investigation based on a three-dimensional KMC model extending the approach by Quiroga and Franco [94].

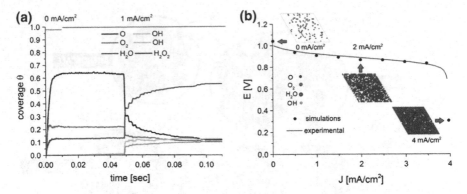

Fig. 29 **a** Calculated response of the ORR species coverage evolution following an applied current step; **b** calculated and experimental polarization curve and associated calculated snapshots of the catalyst surface adspecies coverage (adapted from Ref. [11])

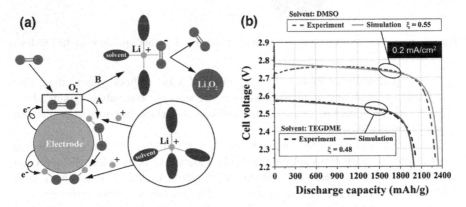

Fig. 30 **a** Schematics of the so-called solution phase and thin-film ORR mechanisms; **b** simulated discharge curves for two solvents with the multiscale cell model detailed in Ref. [95]

4 Conclusions and Open Challenges

The optimization of the performance and durability of batteries and fuel cells can be imagined as the optimization of a mathematical merit function (Fig. 31 for the case of LABs), determined by three factors:

- the intrinsic capacity of the storage materials (batteries) or the intrinsic activity of the catalyst materials (fuel cells);
- their statistical utilization in the porous electrode;
- the macroscopic cell design.

Within this sense, integrative multiscale modeling tools have a crucial role to play for fundamental understanding, diagnostics and design of new electrochemical

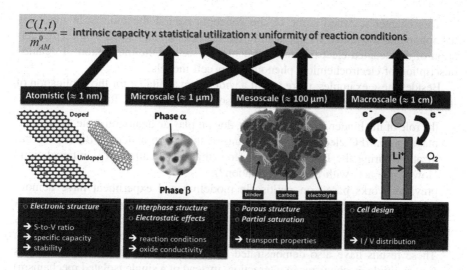

$$\frac{C(I,t)}{m_{AM}^0} = \text{intrinsic capacity} \times \text{statistical utilization} \times \text{uniformity of reaction conditions}$$

Fig. 31 Merit function between scales determining the LAB capacity and cyclability (from Ref. [66])

Fig. 32 Bringing electrochemical energy technologies from the lab to the market (adapted from Ref. [96])

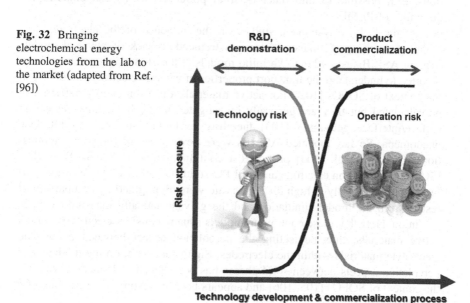

materials and operation conditions. Deep insight based on the modeling of the materials behavior and aging will advise us how these components with optimal specifications could be made and how they can be integrated into operating devices: this can definitely help on boosting the integration of concepts developed at the lab scale into the market (Fig. 32).

This chapter revisited some of the concepts behind the multiscale modeling discipline and illustrated its application through examples related to the consideration of the detailed electrode microstructure as well as the integration of detailed description of electrochemical phenomena in cell models.

Besides these examples, there is still a wide room available for the application of these types of models to help with still unsolved R&D issues, for instance:

– the role of the binder in standard LIBs, does it play an analogous role as Nafion® plays in PEMFC electrodes? (poisoning of the active material? Binder reorganization during the LIB operation? How this binder affects Li^+ intercalation? How it interacts with the SEI formation?);
– previous works based on multiscale modeling and experiment have demonstrated that in PEMFCs the external pollutant CO (introduced in the hydrogen feed) can be used, under some specific operation conditions, as a mitigation factor of the cathode carbon corrosion and the membrane degradation [97–101]. These results have also demonstrated the importance of simulating processes (i.e., multiple mechanisms in interaction, instead of a single isolated mechanism) in order to open the horizon for the discovery of innovative operation conditions. Is it possible to find other external pollutants which can enhance the durability of PEMFCs?;
– besides the lack of systematic studies on the influence of the electrodes formulation on the performance, there is a tremendous lack of physical models treating ASLIBs as a whole. Ab initio models at the atomistic scale have been reported to understand the transport properties of ceramic electrolytes, mainly in the context of SOFCs [102]. For active materials, diffusion energy barriers can be calculated but they are generally investigated in the bulk, thus neglecting surface/interface activation [103] therefore excluding surface and interface phenomena. So far, modeled ASLIBs were devoted to microdevices applications: Danilov et al. [104] proposed a continuum model of ASLIBs with a $LiCoO_2$ intercalation electrode and an LPO solid-state electrolyte. The transport limitations, especially at high discharge rate were highlighted by comparison of results from the model simulation and the galvanostatically measured voltage evolution. Note that these continuum models usually consider electrodes as bulk active materials, thus neglecting the percolation aspect between active and electrolyte materials within the electrodes. Significant work on the modeling of percolation aspects between solid phases has been developed since 20 years in the context of SOFCs [105, 106] and appears here interesting to be extended for ASLIBs.

Multiscale modeling methodologies start to be strongly consolidated now in the electrochemical engineering community but still need significant improvements for their stronger impact on the design and optimization of the next generation of electrochemical power generators. For instance,

– a remaining scientific challenge, is using multiscale physical models in order to get mathematically reduced descriptions which can be used for diagnosis and prognosis. The challenge arises from the strong coupling, synergies, and competitions between the materials and components degradation mechanisms happening at multiple spatiotemporal scales: degradation mechanisms impact on the overall performance decay is not additive. Some degradation mechanisms can cancel with each other, some others can be reinforced. The Bond Graph approach may be interesting for succeeding on deriving reduced mathematical models from multiscale physical models, and still capturing the main physical phenomena determining the cell performance decay;

- usually models need parameters fitting, through the use of specific algorithms to optimize the parameter values in order to fit the calculated observables (e.g., U, I) to experimental data. This process is done at a fixed set of physical assumptions associated to a model. Future needs to include the development of automata algorithms developing, testing, comparing with experimental data and optimizing, multiple (evolutive) generations of models. In such a type of automata, a master routine automatically changing the physics of a model and then subroutines optimize the parameters values in each model generation (Fig. 33);
- stronger modeling efforts are needed in cloud computing favoring the exchange of data, models, and algorithms between groups all around the world (Fig. 33);
- the development of three-dimensional immersive visualization tools is also crucial to allow a better exploitation and understanding of the simulated three-dimensional electrode structures (Fig. 33). This has strong potential to change the paradigm in communicating results, and to ease the communication

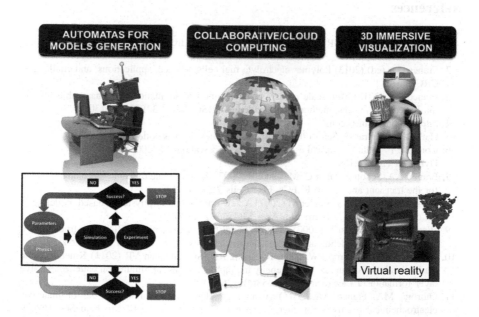

Fig. 33 Schematics of some remaining methodological challenges in the multiscale modeling field

between research and industry as well as in education. A joint project on this aspect applied to electrochemistry is currently ongoing between LRCS in Amiens and UTC in Compiègne (France) and will be the subject of a future publication [107].

Finally, let us remind that history teaches us that energy forms can be used to make practical things without knowing the nature or fully understanding the fundamentals of those energy forms. This affirmation is not unrealistic: Volta invented the first electric battery in 1800, almost 100 years before than electrons were discovered (credited to Thomson in 1897) [1]. Therefore, the subsequent relatively fast penetration of rechargeable batteries in the market was mainly based on trial-error synthesis of materials and testing, and it is only since recent years supported by deep theoretical tools at the single material level. More generally than multiscale modeling, theoretical physics has a very significant role to play in order to one day discover new forms of energy in our Universe or to design ways to manipulate some others that we already know are somewhere there (e.g., neutrinos, dark energy, etc.) to do practical things for the prosperity of the humanity: a completely new type of multiscale modeling tools devoted to the optimization and design of these emerging new technologies will then emerge, and will need to be supported on new physical ingredients and techniques (i.e., going beyond the manipulation of electrons which is the pillar of DFT) even not imagined yet.

References

1. Franco A (ed) (2015) Rechargeable lithium batteries: from fundamentals to applications. Elsevier
2. Franco AA (ed) (2013) Polymer electrolyte fuel cells: science, applications, and challenges. CRC Press
3. Franco AA (2013) Multiscale modelling and numerical simulation of rechargeable lithium ion batteries: concepts, methods and challenges. RSC Adv 3(32):13027–13058
4. http://mobiiarttshirts.com/
5. Haddad L, Duprat G, Seuil (2006) Mondes: Mythes et Images de l'Univers
6. Couenne F, Jallut C, Maschke B, Breedveld P, Tayakout M (2006) Math Comput Modell Dyn Syst 12(2–3):159
7. Franco AA, Schott P, Jallut C, Maschke B (2007) A multi-scale dynamic mechanistic model for the transient analysis of PEFCs. Fuel Cells 7(2):99–117
8. Bessler WG, Gewies S, Vogler M (2007) A new framework for physically based modeling of solid oxide fuel cells. Electrochim Acta 53(4):1782–1800
9. Wang CY, Srinivasan V (2002) Computational battery dynamics (CBD)—electrochemical/thermal coupled modeling and multi-scale modeling. J Power Sources 110(2):364–376
10. Methekar RN, Northrop PW, Chen K, Braatz RD, Subramanian VR (2011) Kinetic Monte Carlo simulation of surface heterogeneity in graphite anodes for lithium-ion batteries: passive layer formation. J Electrochem Soc 158(4):A363–A370
11. Quiroga MA, Franco AA (2015) A multi-paradigm computational model of materials electrochemical reactivity for energy conversion and storage. J Electrochem Soc 162(7): E73–E83

12. Andreaus B, Maillard F, Kocylo J, Savinova ER, Eikerling M (2006) Kinetic modeling of COad monolayer oxidation on carbon-supported platinum nanoparticles. J Phys Chem B 110 (42):21028–21040

13. De Morais RF, Sautet P, Loffreda D, Franco AA (2011) A multiscale theoretical methodology for the calculation of electrochemical observables from ab initio data: application to the oxygen reduction reaction in a Pt (111)-based polymer electrolyte membrane fuel cell. Electrochim Acta 56(28):10842–10856

14. Fantauzzi D, Zhu T, Mueller JE, Filot IA, Hensen EJ, Jacob T (2015) Microkinetic modeling of the oxygen reduction reaction at the Pt (111)/gas interface. Catal Lett 145(1):451–457

15. Eikerling MH, Malek K, Wang Q (2008) Catalyst layer modeling: structure, properties and performance. In: PEM fuel cell electrocatalysts and catalyst layers. Springer, London, pp 381–446

16. Franco AA, Frayret C (2014) Modeling in the design of batteries for large and medium-scale energy storage, book chapter. In: Menictas C, Skyllas-Kazacos M, Lim TM (eds) Advances in batteries for large- and medium-scale energy storage. Elsevier/Woodhead, Cambridge

17. Sheppard D, Terrell R, Henkelman G (2008) Optimization methods for finding minimum energy paths. J Chem Phys 128(13):134106

18. Henkelman G, Uberuaga BP, Jónsson H (2000) A climbing image nudged elastic band method for finding saddle points and minimum energy paths. J Chem Phys 113(22):9901–9904

19. Henkelman G, Jónsson H (2000) Improved tangent estimate in the nudged elastic band method for finding minimum energy paths and saddle points. J Chem Phys 113(22):9978–9985

20. Ferreira de Morais R, Franco AA, Sautet P, Loffreda D (2015) Interplay between reaction mechanism and hydroxyl species for water formation on Pt (111). ACS Catal 5(2):1068–1077

21. Yu Y, Zuo Y, Zuo C, Liu X, Liu Z (2014) A hierarchical multiscale model for microfluidic fuel cells with porous electrodes. Electrochim Acta 116:237–243

22. Malek K, Franco AA (2011) Microstructure-based modeling of aging mechanisms in catalyst layers of polymer electrolyte fuel cells. J Phys Chem B 115(25):8088–8101

23. Kim GH, Smith K, Lee KJ, Santhanagopalan S, Pesaran A (2011) Multi-domain modeling of lithium-ion batteries encompassing multi-physics in varied length scales. J Electrochem Soc 158(8):A955–A969

24. Prigogine I (1967) Introduction to thermodynamics of irreversible processes, vol 1, 3rd edn. Interscience, New York

25. Georgiadis MC, Myrian S, Efstratios N, Gani R (2002) Comput Chem Eng 26:735

26. Franco AA (2005) A physical multi-scale model of the electrochemical dynamics in a polymer electrolyte fuel cell—an infinite dimensional Bond Graph approach. PhD Thesis Université Claude Bernard Lyon-1 (France) no. 2005LYO10239

27. Bozic S, Kondov I (2012). In: Cunningham P, Cunningham M (eds) eChallenges e-2012 conference proceedings. IIMC International Information Management Corporation

28. http://www.knime.org/

29. http://www.unicore.eu/index.php

30. Elwasif WR, Bernholdt DE, Pannala S, Allu S, Foley SS (2012) 2012 IEEE 15th international conference on computational science and engineering, cse, pp 102–110

31. Franco AA (2014) Physical modeling and numerical simulation of direct alcohol fuel cells. In Direct alcohol fuel cells. Springer, Netherlands, pp 271–319

32. Hu G, Li G, Zheng Y, Zhang Z, Xu Y (2014) J Energy Inst 87:163

33. Song D, Wang Q, Liu Z, Eikerling M, Xie Z, Navessin T, Holdcroft S (2005). A method for optimizing distributions of Nafion and Pt in cathode catalyst layers of PEM fuel cells. Electrochimica Acta 50(16):3347–3358

34. Moore M, Wardlaw P, Dobson P, Boisvert JJ, Putz A, Spiteri RJ, Secanell M (2014) Understanding the effect of kinetic and mass transport processes in cathode agglomerates. J Electrochem Soc 161(8):E3125–E3137

35. Xing L et al (2014) Numerical investigation of the optimal Nafion® ionomer content in cathode catalyst layer: An agglomerate two-phase flow modelling. Int J Hydr En 39:9087–9104
36. Dargaville S, Farrell TW (2010) Predicting active material utilization in LiFePO4 electrodes using a multiscale mathematical model. J Electrochem Soc 157(7):A830–A840
37. Zhang X, Ostadi H, Jiang K, Chen R (2014) Reliability of the spherical agglomerate models for catalyst layer in polymer electrolyte membrane fuel cells. Electrochimica Acta 133:475–483
38. Roberts SA, Brunini VE, Long KN, Grillet AM (2014) A framework for three-dimensional mesoscale modeling of anisotropic swelling and mechanical deformation in lithium-ion electrodes. J Electrochem Soc 161(11):F3052–F3059
39. Kriston A, Pfrang A, Popov BN, Boon-Brett L (2014) Development of a full layer pore-scale model for the simulation of electro-active material used in power sources. J Electrochem Soc 161(8):E3235–E3247
40. Liu Z, Battaglia V, Mukherjee PP (2014) Mesoscale elucidation of the influence of mixing sequence in electrode processing. Langmuir 30(50):15102–15113
41. Sanyal J, Goldin GM, Zhu H, Kee RJ (2010) A particle-based model for predicting the effective conductivities of composite electrodes. J Power Sources 195(19):6671–6679
42. Gawel DA, Pharoah JG, Beale SB (2015) Development of a SOFC performance model to analyze the powder to power performance of electrode microstructures. ECS Trans 68 (1):1979–1987
43. Kong W, Zhang Q, Gao X, Zhang J, Chen D, Su S (2015) A method for predicting the tortuosity of pore phase in solid oxide fuel cells electrode. Int J Electrochem Sci 10:5800–5811
44. Bertei A, Pharoah JG, Gawel DAW, Nicolella C (2014) A particle-based model for effective properties in infiltrated solid oxide fuel cell electrodes. J Electrochem Soc 161(12):F1243–F1253
45. Grew KN, Chiu WK (2012) A review of modeling and simulation techniques across the length scales for the solid oxide fuel cell. J Power Sources 199:1–13
46. Cai Q, Adjiman CS, Brandon NP (2011) Modelling the 3D microstructure and performance of solid oxide fuel cell electrodes: computational parameters. Electrochim Acta 56(16):5804–5814
47. Armand M, Tarascon JM (2008) Building better batteries. Nature 451(7179):652–657
48. Kamaya N et al (2011) A lithium superionic conductor. Nat Mater 10:682
49. Huang H (2014) MSc. thesis, Université de Picardie Jules Verne, Amiens, France
50. Cundall PA, Strack ODL (1979) A distinct element method for modeling granular assemblies. Geotechnique 29:47–65
51. Nishida Y, Itoh S (2011) A modeling study of porous composite microstructures for solid oxide fuel cell anodes. Electrochimica Acta 56(7):2792–2800.
52. Nguyen TK, Huang H, Franco AA (2015) Paper in preparation
53. Malek K, Eikerling M, Wang Q, Navessin T, Liu Z (2007) Self-organization in catalyst layers of polymer electrolyte fuel cells. J Phys Chem C 111(36):13627–13634
54. Malek K, Mashio T, Eikerling M (2011) Microstructure of catalyst layers in PEM fuel cells redefined: a computational approach. Electrocatalysis 2(2):141–157
55. Cheng CH, Malek K, Djilali N (2008) The effect of Pt cluster size on micro-morphology of PEMFC catalyst layers-a molecular dynamics simulation. ECS Trans 16(2):1405–1411
56. Franco AA, Gerard M (2008) Multiscale model of carbon corrosion in a PEFC: coupling with electrocatalysis and impact on performance degradation. J Electrochem Soc 155(4):B367–B384
57. Franco AA, Passot S, Fugier P, Anglade C, Billy E, Guétaz L, Fugier P, Mailley S (2009) PtxCoy catalysts degradation in PEFC environments: mechanistic insights I. multiscale modeling. J Electrochem Soc 156(3):B410–B424
58. Coulon R, Bessler W, Franco AA (2010) Modeling chemical degradation of a polymer electrolyte membrane and its impact on fuel cell performance. ECS Trans 25(35):259–273
59. Franco AA, Coulon R, de Morais RF, Cheah SK, Kachmar A, Gabriel MA (2009) Multi-scale modeling-based prediction of PEM Fuel Cells MEA durability under automotive operating conditions. ECS Trans 25(1):65–79

60. Cheah SK, Sicardy O, Marinova M, Guetaz L, Lemaire O, Gélin P, Franco AA (2011) CO impact on the stability properties of PtxCoy nanoparticles in PEM fuel cell anodes: mechanistic insights. J Electrochem Soc 158(11):B1358–B1367
61. Franco AA (2007) Transient multi-scale modelling of ageing mechanisms in a polymer electrolyte fuel cell: an irreversible thermodynamics approach. ECS Trans 6(10):1–23
62. Franco AA, Tembely M (2007) Transient multiscale modeling of aging mechanisms in a PEFC cathode. J Electrochem Soc 154(7):B712–B723
63. Franco AA (2012) PEMFC degradation modeling and analysis, book chapter. In: Hartnig C, Roth C (eds) Polymer electrolyte membrane and direct methanol fuel cell technology (PEMFCs and DMFCs)—Vol 1: fundamentals and performance. Woodhead, Cambridge
64. Quiroga MA, Malek K, Franco AA (2015) J Electrochem Soc, in press.
65. Strahl S, Husar A, Franco AA (2014) Electrode structure effects on the performance of open-cathode proton exchange membrane fuel cells: a multiscale modeling approach. Int J Hydrogen Energy 39(18):9752–9767
66. Franco AA, Xue KH (2013) Carbon-based electrodes for lithium air batteries: scientific and technological challenges from a modeling perspective. ECS J Solid State Sc Tech 2(10): M3084–M3100
67. Xue KH, Nguyen TK, Franco AA (2014) Impact of the cathode microstructure on the discharge performance of lithium air batteries: a multiscale model. J Electrochem Soc 161(8): E3028–E3035
68. Bevara V, Andrei P (2014) Changing the cathode microstructure to improve the capacity of Li-air batteries: theoretical predictions. J Electrochem Soc 161(14):A2068–A2079
69. Olivares-Marín M, Palomino P, Enciso E, Tonti D (2014) Simple method to relate experimental pore size distribution and discharge capacity in cathodes for Li/O2 batteries. J Phys Chem C 118(36):20772–20783
70. Weber AZ, Mench MM, Meyers JP, Ross PN, Gostick JT, Liu Q (2011) Redox flow batteries: a review. J Appl Electrochem 41(10):1137–1164
71. Duduta M et al (2011) Semi-Solid lithium rechargeable flow battery. Adv Energy Mater, 1 (4):511–516
72. Hamelet S et al (2012) Non-aqueous Li-based redox flow batteries. J Electrochem Soc 159 (8):A1360–A1367
73. Hamelet S, Larcher D, Dupont L, Tarascon JM (2013) Silicon-based non aqueous anolyte for li redox-flow batteries. J Electrochem Soc 160(3):A516–A520.
74. Brunini VE, Chiang YM, Carter WC (2012) Modeling the hydrodynamic and electrochemical efficiency of semi-solid flow batteries. Electrochimica Acta 69:301–307
75. Smith KC, Chiang YM, Carter WC (2014) Maximizing energetic efficiency in flow batteries utilizing non-Newtonian fluids. J Electrochem Soc 161(4):A486–A496
76. Grinberg L, Karniadakis G, Insley JA, Papka ME Brown University and Argonne National Laboratory. https://www.youtube.com/watch?v=tBga86M9Gm4 and https://www.youtube.com/watch?v=0hibGZi8TWs
77. Franco AA et al (2015) WONDERFUL project (Conseil Régional de Picardie and European Regional Development Fund)
78. Hoogerbrugge PJ, Koelman JMVA (1992) Simulating microscopic hydrodynamic phenomena with dissipative particle dynamics. Europhys Lett 19(3):155
79. Darling RM, Meyers JP (2003) Kinetic model of platinum dissolution in PEMFCs. J Electrochem Soc 150(11):A1523–A1527
80. Fowler MW, Mann RF, Amphlett JC, Peppley BA, Roberge PR (2002) Incorporation of voltage degradation into a generalised steady state electrochemical model for a PEM fuel cell. J Power Sources 106(1):274–283
81. Baxter SF, Battaglia VS, White RE (1999) Methanol fuel cell model: anode. J Electrochem Soc 146(2):437–447

82. Bao C, Bessler WG (2015) Two-dimensional modeling of a polymer electrolyte membrane fuel cell with long flow channel. Part I. model development. J Power Sources 275:922–934

83. Zhang H, Haas H, Hu J, Kundu S, Davis M, Chuy C (2013) The impact of potential cycling on PEMFC durability. J Electrochem Soc 160(8):F840–F847

84. Biesheuvel PM, Franco AA, Bazant MZ (2009) Diffuse charge effects in fuel cell membranes. J Electrochem Soc 156(2):B225–B233

85. Franco AA, Schott P, Jallut C, Maschke B (2006) A dynamic mechanistic model of an electrochemical interface. J Electrochem Soc 153(6):A1053–A1061

86. Quiroga MA, Xue KH, Nguyen TK, Tułodziecki M, Huang H, Franco AA (2014) A multiscale model of electrochemical double layers in energy conversion and storage devices. J Electrochem Soc 161(8):E3302–E3310

87. Damasceno Borges D, Franco AA, Malek K, Gebel G, Mossa S (2013) Inhomogeneous transport in model hydrated polymer electrolyte supported ultrathin films. ACS Nano 7 (8):6767–6773

88. Borges DD (2013) Etude computationnelle de la formation d'un film ultra-mince de Nafion à l'intérieur d'une couche catalytique de PEMFC. Doctoral dissertation, Université de Grenoble

89. Damasceno Borges D, Gebel G, Franco AA, Malek K, Mossa S (2014) Morphology of supported polymer electrolyte ultrathin films: a numerical study. J Phys Chem C 119 (2):1201–1216

90. Van der Ven A, Ceder G (2000) Lithium diffusion in layered Li x CoO$_2$. Electrochem Solid-State Lett 3(7):301–304

91. Zhdanov VP (2007) Simulations of processes related to H$_2$–O$_2$ PEM fuel cells. J Electroanal Chem 607(1):17–24

92. Zhdanov VP, Kasemo B (2003) Role of the field fluctuations in electrochemical reactions. Appl Surf Sci 219(3):256–263

93. www.modeling-electrochemistry.com

94. Blanquer G, Yin Y, Quiroga M, Franco AA (2015) J Electrochem Soc (Submitted)

95. Xue KH, McTurk E, Johnson L, Bruce PG, Franco AA (2015) A comprehensive model for non-aqueous lithium air batteries involving different reaction mechanisms. J Electrochem Soc 162(4):A614–A621

96. Malek K, Maine E, McCarthy IP (2014) A typology of clean technology commercialization accelerators. J Eng Tech Manage 32:26–39

97. Franco AA, Guinard M, Barthe B, Lemaire O (2009) Impact of carbon monoxide on PEFC catalyst carbon support degradation under current-cycled operating conditions. Electrochim Acta 54(22):5267–5279

98. Parry V, Berthomé G, Joud JC, Lemaire O, Franco AA (2011) XPS investigations of the proton exchange membrane fuel cell active layers aging: characterization of the mitigating role of an anodic CO contamination on cathode degradation. J Power Sources, 196(5):2530–2538

99. Engl T, Käse J, Gubler L, Schmidt TJ (2014) On the positive effect of CO during start/stop in high-temperature polymer electrolyte fuel cells. ECS Electrochem Lett 3(7):F47–F49

100. Franco A, Lemaire O, Escribano S (2014) US Patent 8,871,399. Washington, DC: US Patent and Trademark Office.

101. Franco AA, Lemaire O, Escribano S (2008) FR patent EN 08. 50875

102. Islam MS, Fisher CA (2014) Lithium and sodium battery cathode materials: computational insights into voltage, diffusion and nanostructural properties. Chem Soc Rev 43(1):185–204

103. Morgan D, Van der Ven A, Ceder G (2004) Li conductivity in Li x MPO 4 (M = Mn, Fe, Co, Ni) olivine materials. Electrochem Solid-State Lett 7(2):A30–A32

104. Danilov D, Niessen RAH, Notten PHL (2011) Modeling all-solid-state Li-ion batteries. J Electrochem Soc 158(3):A215–A222

105. Costamagna P, Costa P, Arato E (1998) Some more considerations on the optimization of cermet solid oxide fuel cell electrodes. Electrochimica Acta 43(8):967–972

106. Bertei A, Nucci B, Nicolella C (2013) Microstructural modeling for prediction of transport properties and electrochemical performance in SOFC composite electrodes. Chem Eng Sci 101:175–190
107. Franco AA, Thouvenin I et al (2014) MASTERS project (Conseil Régional de Picardie and European Regional Development Fund)

Cost Modeling and Valuation of Grid-Scale Electrochemical Energy Storage Technologies

Kourosh Malek and Jatin Nathwani

Abstract Electrochemical Energy storage (ES) technologies are seen as valuable flexibility assets with their capabilities to control grid power intermittency or power quality services in generation, transmission & distribution, and end-user consumption side. Grid-scale storage technologies can contribute significantly to enhance asset utilization rate and reliability of the power systems. The latter is particularly critical for deployment of regional and national energy policies of implementing renewable sources. Once the suitable storage technology is chosen, modeling and simulation of electrochemical storage devices are utilized extensively for performance or life cycle prediction purposes. The main challenge of adopting electrochemical storage technologies among utilities is how to match the right energy storage technology for a site-specific grid configuration to an appropriate grid service. The majority of system-level modeling efforts do not provide information that can be used for valuation of storage technologies. Battery performance models generally suffer from lacking techno-economic predictions and accurate assessment of performance characteristics of the emerging ES technologies. This chapter introduces a valuation framework that is built upon high-level electrochemical storage models. This valuation model can characterize and quantify different grid applications and services for which electrochemical storage devices are used. Taking local differences in electricity markets and storage value for several grid applications and services, the modeling framework is employed in case studies to identify the value that storage systems can provide to the grid.

K. Malek (✉) · J. Nathwani
Department of Management Sciences, University of Waterloo, Waterloo, ON, Canada
e-mail: kourosh.malek@nrc.gc.ca

K. Malek
NRC-EME, 4250, Wesbrook Mall, Vancouver, BC V6T 1W5, Canada

© Springer-Verlag London 2016
A.A. Franco et al. (eds.), *Physical Multiscale Modeling and Numerical Simulation of Electrochemical Devices for Energy Conversion and Storage,*
Green Energy and Technology, DOI 10.1007/978-1-4471-5677-2_7

1 Introduction

The electricity grid is an essential regional asset that provides infrastructure for local electrical energy demand or export markets. In recent years, electricity distribution networks are encountered considerable challenges, such as aging network assets, the installation of new distributed generators, carbon reduction obligations, implementing regulatory incentives, and the capability of adopting new technologies for electricity generation, transmission, and distribution [1, 2]. There is a recent trend in which the energy industry is transformed toward producing a more sustainable production of electricity. In many countries, including Canada, grid capital assets are coming close to the end of life as they are not able to satisfy increasing demand conditions. In particular, increasing use of intermittent renewable energy generation can create new challenges for grid stability and reliability. By 2035, renewable sources, such as wind and PhotoVoltaics (PV), will account for nearly half of the increase in global power generation [3]. The increasing share of renewable sources in the global power market can also create challenges in the power sector such as investment risks and supply reliability [3].

Energy storage (ES) technologies with their capabilities to control power intermittency, can provide various services along the electricity value chain at generation, transmission and distribution (T&D), retail, and end-user consumption. Examples of these services are energy or power arbitrage, backup power, frequency regulation, peak shaving, and power reliability. The role of storage technologies is to transform electricity into a different form of energy (e.g., chemical, potential, or mechanical), store the energy for certain periods of time (from seconds to days), and recover electrical energy in case of needs [4]. Despite the fact that by focusing on the only one application, ES systems increase the operational cost of the distributed electricity system [5–8], ES technologies can play a vital role in reducing the overall upgrade cost of the electricity grids in the presence of renewable sources.

The main challenge of adopting ES technologies is how to match the right technology to appropriate grid service for a site-specific configuration. There are numerous technical assessment and engineering tools that provide substantial information around technical value of storage technologies. The tools are usually built around electricity production or transmission reliability models, with no or little market and financial-driven information [6]. The majority of the tools thus suffer from lacking technology management and business information, making them difficult to be used by managers for decision making purposes. In order to address the gaps, we introduce relevant frameworks from business and technology management discipliners that can be used for valuation and early adoption of grid-scale emerging storage technologies. Such analysis approaches integrate the technical database into business opportunity assessment with additional features that enable fast screening of the emerging technologies. On a general basis, these concepts form the basis of a unique management methodology to assess alternative technology solutions and provide unbiased reliable information upon which reliable management decisions for adopting new technologies can be made.

2 Methodology

Customized for grid-scale storage technologies, our analysis methodology stays on the basis that any storage deployment is identified by key characteristics that include location, grid application or services (e.g., backup, grid reliability, frequency regulation, arbitrage), type of electricity market (e.g., regulated vs. deregulated), type of ownership (utility owned vs. privately owned), and type of ES technology to be deployed (e.g., performance, time of discharge, response time). Several valuation tools have been developed to analyze the value of distributed or bulk storage technologies for various grid applications [9]. The underlying assumption in the majority of those tools is that the storage system will not significantly influence market conditions, and therefore, existing market prices are used as the input market parameters [9]. There is a fundamental difference between such valuation tools and those of electricity production cost models, where an extensive system operation and knowledge of economic dispatch is required for the latter. Our focus in this proposal is, however, entirely on the former class of valuation tools.

Among the most common valuation approaches and tools that have been widely utilized by utilities and independent consultant are NREL valuation (an analysis tool to evaluate the operational benefit of commercial storage, including load-leveling, spinning reserves, and regulation reserves) [10, 20]; Energy Storage Valuation Tool (ESVT) developed by EPRI [21] has proposed a methodology for separating and clarifying analytical stages for storage valuation. ESVT calculates the value of ES by considering the full scope of the electricity system, including system/market, transmission, distribution, and customer services; and ES-Select™ designed and developed by DNV-KEMA [11]. In ES-Select™, the user needs to choose where ES is connected to an electric grid [11] and the emphasis is more on "simplicity" and user-friendly functionalities for screening and educational purposes than ultimate "accuracy" [11]. Therefore, inputs are assumed by default or entered by the user in a certain range of accuracy.

According to a recent report by Navigant [12], among the main shortcomings with existing ES valuation tools are a lack of standardization among valuation model and limits on the data available on storage technologies. To the best of our knowledge, none of the existing valuation tools compile or utilize information about business models. Storage ownership is often used as a mean to identify the business model and has been used to define the group of stakeholders that retain the profit or losses of storage asset. Ownership can be with private individuals, utilities, or "Gentailers." The latter is a new type of ownership in which the retailer has the option to install a storage system, and therefore reduces the supply cost of the customers [13]. ES-select tool [11] was utilized as a framework to quantify the feasibility and reliability of the ES systems.

3 Performance Matrix

In order to fully assess value proposition of ES technologies, formulate their risk and opportunity profile, and develop implementation plan, a number of analyses frameworks need to be developed and utilized from techno-economics and business operation perspectives. The underlying idea is to focus on a specific storage technology, and compare it to other similar technologies for grid applications by mapping its technological advantages/disadvantages, and innovation capacity. Here, we particularly focus on technology road mapping, technology development matrix, and technology valuation grid.

Technology development matrix (TDM) is linking market needs to technology attributes to key technical parameters. TDM is another form of technology management framework that can help technology managers and system integrators identify the technical R&D gaps and target suitable market opportunities for adopting their technologies. It translates what consumer wants into technical goals for a given market. When constructed carefully, it forms the technology plan and R&D projects portfolio. When used as a collaborative tool, it brings technical team together in a common goal to address commercialization gaps. However, market needs change, so as the state-of-the-art (SoTA) performance and key underlying assumptions. TDM should be a live document and updated regularly. In reality, the *stage-gate* process that are developed internally in many firms, are normally a workable version of TDM. They serve the initial purpose of understanding the landscape, technology priorities, and making a decision of project's portfolio mix.

Storage performance matrix is an integral part of TDM for ES technologies that describes the acceptable range of technical attributes for a given grid service. A brief description of storage performance matrix is provided here by concentrating on the application of TDM for technology mapping of the grid-scale ES technologies. Based on the types of services and installed capacity, ES technologies in electrical energy systems can be grouped into chemical storage (batteries or hydrogen), potential energy (pumped hydro or compressed air), electrical energy (supercapacitor), mechanical energy (flywheels), and magnetic energy (supermagnetic ES). Storage systems include a number of technologies in different Technology readiness levels (TRLs). The performance matrix that characterizes and compares different technologies are separated from the location and services that they can provide. Other categorizations are based on the time of use (TOU), short-term, long-term, and distributed storage, or level of maturity and technology advancement.

The cost and reliability of an ES technology are function of several key factors. Among those factors are round-trip efficiency (the ratio of the released electrical energy to the stored energy), cycle life (the number of times that the device can get

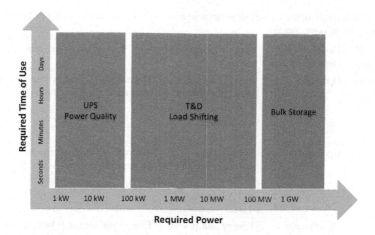

Fig. 1 Required power and response time for different grid-scale storage services

discharged and charged while maintaining a minimum required efficiency), power rating ($/kW), and energy rating ($/kWh). Moreover, capital and operating costs determine economic viability and service profitability, Fig. 1 illustrates required power and response time for different grid-scale storage services.

The real benefit of ES technologies have been studied extensively in different markets (e.g., arbitrage, regulation services, and T&D) [14–17]. By focusing on only one single application, storage technologies has not shown significant value and service profitability [3]. The reason is that the actual choice of appropriate storage technology for a specific grid application is the interplay between time of usage, charge/discharge time, and cost that may not collectively lead to a profitable operation for a single storage technology or in a single application. Commercial viability requirements and cost effectiveness of storage solutions for grid applications is still under debates in academic and business-management literature [18]. Figure 3 captures the characteristic time and cost benefit data for specific application and maps some storage technologies. As indicated in various studies, no single ES system can provide multiple grid application requirements [11]. Moreover, some storage technologies may complement each other for multiple services, where combining services could lead to cost recovery and profitability in the long-run [3, 6]. A performance matrix is the basis of the ES valuation which characterizes a storage technology for various applications in electricity grid systems. The most common attributes in the metrics are provided in Table 1. This is an example of TDM in which elements of storage performance matrix and system attributes are described for different storage technologies, both at system and standalone technology levels.

Table 1 An example of technology development matrix with selected elements from performance matrix and the linkages therein

TDM level of attribute	Category of element/attribute	Performance matrix element	Brief description of the element
Technology	Operation	Energy storage capacity [kWh or Ah]	The amount of energy that can be recovered at a given time
		SoTA versus Target	
	Operation	Charge and discharge rates [kW or A]	The rate at which energy is consumed or stored in a storage system
		SoTA versus Target	
	Performance	Energy and power density [kWh/m^3 or kWh/ton]	Energy per weight [kWh/ton] or energy per volume [kWh/m^3] are considered as energy and power factors
		SoTA versus Target	
System	Performance	Round-trip efficiency [%]	The percentage of the additional required energy during charging is expressed as round-trip efficiency [%]
		SoTA versus Target	
	Cost	Levelized Cost of Storage [$/kW]	The levelized cost of energy storage (LCOES) is defined as the overall cost of ownership of storage over the investment period divided to the total delivered energy in that period
		SoTA versus Target	
	Durability	Lifetime [cycles, years, kWh$_{life}$]	The lifetime of a storage system can be measured by the number of charge/discharge cycles at given energy capacity
		SoTA versus Target	

4 Techno-Economic Cost Modeling

This section describes a cost model and valuation methodology for calculating the benefits of grid-scale ES technologies. The background, methodology, assumptions, and detailed necessary steps are provided in order to build a computational strategy for assessing the value of certain grid services which maximize the benefits of a given ES asset.

4.1 Analytics Framework

In our valuation, each deployment is identified by key characteristics that include location, type of market, type of ownership and type of ES technology to be deployed. The database in ES-Select™ [11] are initially utilized for the majority of technical attributes, applications, and cost data, including installation costs. The first layer is common among exiting valuation tools, in which the monetary value of a specific storage technology (or a group of technologies) for a given grid-service application (or a group of multiple services) is calculated based on input financial information and storage technology attributes. Several databases are required in this layer to determine which storage technology can fulfill the technical requirement of certain applications on the grid. The output of this layer is a feasible subset (binary) of applications for a given storage technology or a subset of storage technologies which are feasible for a given grid service. Finally, using input financial and asset ownership information, one can calculate the economic value of each benefit. Section 4.2 provides the basic equations and relationships used to calculate the asset benefits.

The second layer of the logic model utilizes ownership type and market structure to determine which business model can fulfill the monetary value of the benefits calculated within the first layer for each binary choice of (storage, application). Each business model is described by a series of characteristics related to market structure, asset ownership, and range of risk profile, benefit, and asset location. The algorithm in the second layer will utilize the feasibility score of a given business model for a given storage-application combination.

For the sake of simplicity, only primary applications are considered in our model. Thus, we do not consider the benefit of a given storage technology for a bundled set of applications same as that considered in ES-Select™ and other valuation tools [11]. However, the benefit stacking for multiple applications can be determined separately from each [storage, application] binaries.

4.2 Determining Storage Benefits

In order to evaluate the financial benefit of a given storage technology, one needs to determine the type of storage asset, the application (grid service) that storage asset provides, owner of the storage asset, the type of market that storage asset will be deployed in, and location of the asset in the electricity grid. A grid application describes how the storage system can be utilized for a specific grid service and business model describes how the asset owner can monetize that service to gain certain value or benefit.

Table 2 provides the primary list of grid locations (generation, transmission and distribution-T&D, and end user), technologies, applications, ownership, and market structures that are commonly considered. The choice of storage technologies is

Table 2 List of grid locations, technologies, applications, ownerships, and market structures that are considered in this work

Attributes	Types	Definition	Comments
Location	Generation, transmission and distribution (T&D) and end user	Determines on what part of the grid the storage is located	End user can be further categorized to residential and commercial users
Technology	Lithium ion batteries, (vanadium) redox flow, sodium sulfur (NaS), hydrogen storage, advanced lead acid, and compress air energy storage	Determines the type of energy storage technology	Only those with high TRL and available demonstration data are considered here
Application	Arbitrage, supply capacity, backup, power quality, frequency regulations	Indicates how the storage asset will be used and what kind of service will be provided by the storage	The most common services are considered
Ownership	Utility, Independent Power Producers including nonutility merchant (IPPs), private individuals and end users (End User)	Identify the entity who owns the asset and therefore accept the capital cost, benefit/loss and risk of capital	For the sake of simplicity, only three levels of ownership are considered
Market	Regulated, de-regulated, mixed	Specify the jurisdictions and electricity market for deploying the asset and electricity pricing structure	Each type represents a jurisdiction in Canadian electricity market

based on two distinct factors: Technology readiness (maturity) Level or TRL of the storage technology and the extent of demonstration projects or available real-time data that have utilized those technologies. Based on these factors, lithium ion battery (LiB), redox flow battery (RFB), sodium sulfur (NaS) battery, hydrogen storage, advanced lead acid battery (LAB), and compressed air energy storage (CAES) are chosen as the primary storage technologies. As for applications, our focus is on selected application services, most important of which are energy time-shift (arbitrage), supply capacity, utility backup (Service reliability), power quality, and frequency regulations (firming renewables). Although there is no clear consensus and standard for defining storage services, we refer to the definition in ES-Select™ [11].

Three types of market structures are considered, including highly regulated, deregulated, and mix of regulated and deregulated markets. In a regulated electricity market, utilities incorporate all or most of the services and electricity deliveries are vertically integrated. In a deregulated market, on the other hand, the services are not vertically integrated by utilities. Instead, other independent power producers (IPPs), distributes or other merchant generators are allowed to participate in the electricity market. In the case of mixed regulated-deregulated market structure, the generation side is highly regulated and is managed by utilities, whereas distribution and end-user sides are deregulated. Owners of the storage asset are divided into utilities, a nonutility merchant or an IPP, and private individuals (end users). As storage asset owner, utilities maintain, and operate the transmission line, whereas IPPs deploy the ES asset independently in whole-sale electricity market. Private owners are end users of electricity. Although we have limited our attention to the attributes discussed in Table 2, the concepts and methodologies are scalable and can be extended to a wider range of application services and technologies. In general, one is able to define his or her own technology or application by adjusting these default values.

The benefit of storage is ultimately described by return on the total cost of capital for a specific period of time (asset life time) based on several financial outputs that include net present value (NPV), internal rate of return (IRR), the total cost of ownership (TCO), and cash flow.

Figure 2 provides an overview of the cost components for storage asset. The expected (annual) benefits ($/kW) are simply defined as default in the application database for each application type. Qualitatively, the benefits are ranked as regulation services > system capacity > arbitrage > backup. The annual cost of expenses ($/year/kW) is calculated from the annual cost of operation (C_{ops}) and maintenance (C_m):

$$C_{exp} = (C_{ops} + C_m) \qquad (1)$$

The annual cost of operation is calculated by:

$$C_{ops} = \frac{C_{charge} \times L_{ops}}{1000} \qquad (2)$$

where L_{ops} represents the annual operation loss of the storage is the loss of storage performance and is defined as kWh/year/kW and C_{charge} represents the cost of battery charge. C_m is an input parameter in the storage technology database.

The cost of storage installation C_{SI} is the sum of installation cost C_I and capital cost of storage C_S in $/kW:

$$C_{SI} = (C_I + C_S) \qquad (3)$$

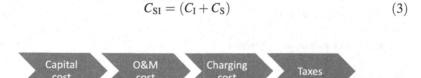

Fig. 2 Overview of cost components for a storage asset

By factoring in the discount rate over asset life time (n) and calculating present value (PV) of the annual cost of expense, one can calculate the TCO as

$$TCO = \left[PV(C_{exp}) + C_{SI} + PV(C_R) \right] \qquad (4)$$

where C_R is the replacement cost. The present value of the annual benefits or PV (B) are calculated using the discounted (interest) rate from the financial database and the annual benefits defined in the application database. The annual net present value of benefits or annual cash flow is calculated by:

$$\text{Cash Flow} = CF = \left[PV(B) - PV(C_{exp}) \right] \qquad (5)$$

Table 3 List of essential input parameters

Input parameter	Unit	Definition	Source
Cost of maintenance per year	$/year/kW	Includes balance of system (BOS)	[11]
Required application discharge duration	Cycles	Technical requirements for specific application	[19]
Annual benefit	$/kW		[11]
Cost of energy used for charging	$/MWh		This work
Cycle life at 10 % DoD	Cycles		This work
Cycle life at 80 % DoD	Cycles		This work
Discount rate	%/year	Also referred to as interest rate	User input
Escalation of benefits	%/year	Projected annual increase of benefits	User input
10-year total benefits	$B		[11]
AC round-trip efficiency	%		[19]
Feasibility score for fulfilling application requirement	%	Scores based on power, energy, frequency of use	[11, this work]
Feasibility score for selected location	%	Scores are different for selected locations on grid	[11, 19]
Feasibility score for maturity	%	Lab-scale, prototype, precommercial or fully commercial (TRL level)	[11, 19] This work
Feasibility score for selected ownership	%		This work
Electricity price escalation	%/year		User input
AC storage cost	$/kW		[11, 19] this work
Storage discharge duration	Cycles	From technology matrix	[11, 19]

The payback year is defined as the year (n) in which the cumulative cash flow at that year is equal to C_{SI}.

$$\sum_{1}^{n} CF = C_{SI} \tag{6}$$

Tax rates (τ) will be included in all cost and benefit terms. One should notice that a single revenue stream (from a single application service) usually does not lead to a short (<10 years) payback time. Only multiple revenue streams could lead to net benefits in a reasonable payback period as illustrated by many studies [19]. Our approach is simplified compared to a more statistical basis that has previously employed in the literature [11]. Note that the effect of electricity price increase is captured by electricity price escalation factor as an input parameter within the financial database. Finally, IRR is calculated as the discounted rate under the assumption that the net cash flow is zero. Table 3 provide the list of essential input parameters. All other parameters not listed in this table are taken as default in the databases but can be adjusted if necessary.

5 Databases

5.1 Database of Storage Technologies

The storage database includes detailed information for different ES technology or types (Performance Metrics). This data has been obtained from several surveys and RFP processes in various jurisdictions. In contrast to a range of accuracy used in ES-select, here we mainly utilize a single, average value for the parameters (Table 4).

Table 4 List of essential parameters in storage technology database [11]

Storage technology	Discharge duration (h)	Specific energy (kWh/ton)	Energy density (kWh/m³)	Round-trip efficiency (at 80 % DoD)	Cycle life at 80 % DoD	Response time
LIB	2.5	100	110	0.885	8000	ms
Advanced lead acid	3.5	24	50	0.85	240	ms
RFB	4	9.5	18	0.63	8000	ms
Sodium sulfur	6.5	110	135	0.765	6000	ms
H₂ storage	6.5	155	120	0.60	10,000	Second
CAES	4			0.65	20,000	Second

Table 5 List of selected attributes for application database [11]

Applications name	Discharge duration (h)	Annual benefit ($/kW)	Total 10-year market potential ($B)	Required response time	Required deep cycles (80 % DoD) (cycles/year)	Market potential 10 years (GW)
Arbitrage	7	100	11	Hours	190	21.34
Supply capacity	6	101	12.1	Hours	190	22.85
Frequency regulations	0.5	560	1.96	Second	4000	3.31
Utility backup	2	330	9.01	Second	100	9.53
Power quality	0.02	150	8.3	ms	500	11.95

5.2 Database of Storage Applications

Despite considerable improvements, there is no consensus in the definition of services that can be given by various storage technologies [19]. The most common services include energy time-shift (arbitrage), power quality, frequency regulations, backup, and supply capacity. The database table includes application name, discharge time, annual benefits, market potential, and the minimum required deep (80 % depth of discharge) cycles (Table 5).

6 Storage Valuation

Key characteristics of storage systems for particular markets in the electricity energy system are illustrated in Table 1, where typical ES applications are characterized in view of different performance attributes. ES market and its associated applications span on a variety of locations along the electricity value chain, Fig. 3. For instance, on the generation side, the addressable market for ES is improving power quality or usage of existing generation sources.

The primary step in valuation of ES technologies for a specific service application is to identify technical parameters (power/energy density, life time, life cycle, cycle ability, and cost) using a ranking strategy for each storage technology based on the various attributes. Figure 4 shows an example of the attributes (L: location; M: maturity level; A: meeting application requirement; C: cost requirement) for NaS, lithium ion (LIB-e) and vanadium redox flow batteries (VRFB), mapped on spider charts for arbitrage as a potential service application. Ranking feasibility scores for this application were obtained for different batteries for a given application area. The charts are obtained from ES-select^{TM} tool [11]. The results have also indicated feasibility order for the above configuration as: NaS > Li-ion > A-VRFB, where A-VRFB stands for the advanced VRFB. The financial indicators such as NPV and

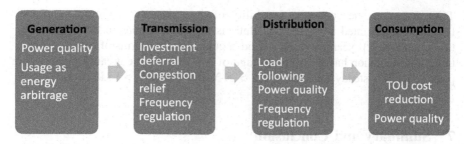

Fig. 3 Energy storages market and their potential applications along electricity value chain

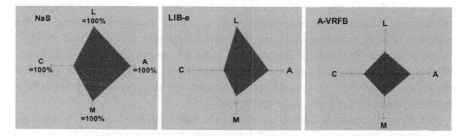

Fig. 4 Ranking feasibility scores for different batteries for a given application. The charts are obtained from ES-select [11]. *L* location; *M* maturity level; *A* meeting application requirement; *C* cost requirement

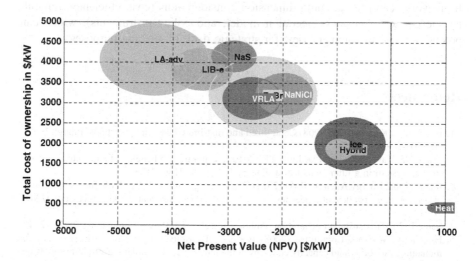

Fig. 5 Total cost of ownership versus NPV ($/kW) for selected storage solutions. The charts are obtained from ES-select [11]. *NaNiCl* sodium nickel chloride; *LIB-e* lithium ion battery; *LA-adv* advanced lead acid; *VRLA* valve regulated lead acid; *NaS* sodium sulfur; ice/heat represents the charge/discharge cycles of a thermal battery

TCO determine the economic feasibility of the storage technologies over their lifetime, as illustrated in Fig. 5. Calculations suggest that none of the battery solutions fulfill the 20 years payback period requirements. In terms of discharge duration, the calculation has shown advantage of A-VRFB for the greatest range where peak demand is steady for 3–6 h (NaS > A-VRFB > VRFB > Li-ion).

7 Summary and Conclusion

Current valuation and technical assessment tools provide substantial information around technology readiness and maturity level of emerging technologies; however, only few of the existing approaches use market driven and business-management information. Technology management tools can help managers evaluate market readiness of new technologies to support new investment decisions and strategic business actions. Technology management tools are essentially different from traditional management and business intelligence in which they provide practical guideline, framework, and modeling techniques to understand and implement business processes for early stage technologies.

We have discussed a bottom-up approach that employs a set of technology management frameworks to support business-management decision of adopting grid-scale storage technologies for grid services and variable electricity generation. Among those technology management tools, severals are employed from matrix management techniques such as TDM, technology oad mapping, and technology valuation grid. For industry looking to adapt new ES technologies, such analysis frameworks can provide multi-dimension considerations (cost, efficiency, reliability, best practice business operation model, and policy instruments), which can potentially lead to complete view for strategic decision making purposes.

References

1. Bouffard F, Kirschen DS (2008) Centralized and distributed electricity systems. Energy Policy 36:4504–4508
2. Wade NS, Taylor PC, Lag PD, Jones PR (2010) Evaluating the benefits of an electrical energy storage system in a future smart grid. Energy Policy 38:7180–7188
3. International Energy Agency (IEA) (2013) World energy outlook. Available from http://www.iea.org/newsroomandevents/speeches/131112_WEO2013_Presentation.pdf. Accessed 10 May 2014
4. Scott J (2004) Distributed generation: embrace the change. Power Engineering 2(18):12–13
5. Electricity Advisory Committee (2008) Bottling electricity: storage as a strategic tool for managing variability and capacity concerns in the modern grid. Available from http://www.oe.energy.gov/final-energy-storage_12-16-08.pdfS. Accessed 19 Aug 2010
6. Sandia National Laboratories (2004) Energy storage benefits and market analysis handbook, SAND 2004-6177

7. Walawalkar R, Apt J (2008) Market analysis of emerging electric energy storage systems, DOE/NETL-2008/1330
8. Yang Z, Zhang J, Kintner-Meyer MCW, Lu X, Choi D, Lemmon JP, Liu J (2011) Electrochemical energy storage for green grid. Chem Rev 211:3577
9. Pearre S, Swan LG (2014) Applied energy, Article in press
10. Sandia National Laboratories (2004) Energy storage benefits and market analysis handbook, SAND 2004-6177
11. Sandia National Laboratories/DNV-KEMA (2014) ES-select documentation and user's manual. Available from http://www.sandia.gov/ess/ESSelectUpdates/ES-Select_Documentation_and_User_Manual-VER_2-2013.pdf. Accessed 10 May 2014
12. Lamontagne C (2014) Navigant survey of models and tools for the stationary energy storage industry. Available from http://www.slideshare.net/navigant/survey-of-models-and-tools-for-the-stationary-energy-storage-industry-february-2014. Accessed 01 Oct 2014
13. Pye J (2014) Three possible business models for distributed storage March 2014. Available from http://www.wattclarity.com.au/2014/03/three-possible-business-models-for-distributed-storage/. Accessed 01 Oct 2014
14. Denholm P, Sioshansi R (2009) The value of compressed air energy storage with wind in transmission-constrained electric power systems. Energy Policy 37:3149–3158
15. Hittinger E, Whitacre JF, Apt J (2012) What properties of grid energy storage are most valuable? J Power Sources 206:436–449
16. Denholm P et al, NREL (2010) The role of energy storage with renewable electricity generation. Available from http://www.nrel.gov/docs/fy10osti/47187.pdf. Accessed 10 May 2014
17. Electric Power Research Institute, Electricity Energy Storage Technology Options (2010) http://my.epri.com/portal/server.pt? Abstract id = 000000000001020676. Accessed 10 May 2014
18. EPRI (2014) The integrated grid: realizing the full value of central and distributed energy resources. Available from http://www.epri.com/abstracts/Pages/ProductAbstract.aspx? ProductId=3002002733. Accessed 10 May 2014
19. EPRI, US Department of Energy (2003) EPRI-DOE handbook of electricity storage for transmission and distribution applications, 1001834
20. Denholm P et al, NREL (2013) The value of energy storage for grid applications. Available from http://www.nrel.gov/docs/fy13osti/58465.pdf. Accessed 10 May 2014
21. Kaun B (2013) Cost-effectiveness of energy storage in California, Application of the EPRI Energy Storage Valuation Tool to Inform the California Public Utility Commission Proceeding R. 10-12-007. Available from http://www.cpuc.ca.gov/NR/rdonlyres/1110403D-85B2-4FDB-B927-5F2EE9507FCA/0/Storage_CostEffectivenessReport_EPRI.pdf. Accessed 10 May 2014

Printed in the United States
By Bookmasters